TOPICS ON STABILITY AND PERIODICITY IN ABSTRACT DIFFERENTIAL EQUATIONS

SERIES ON CONCRETE AND APPLICABLE MATHEMATICS

Published

Vol. 1 Long Time Behaviour of Classical and Quantum Systems
edited by S. Graffi & A. Martinez

Vol. 2 Problems in Probability
by T. M. Mills

Vol. 3 Introduction to Matrix Theory: With Applications to Business and Economics
by F. Szidarovszky & S. Molnár

Vol. 4 Stochastic Models with Applications to Genetics, Cancers, Aids and Other Biomedical Systems
by Tan Wai-Yuan

Vol. 5 Defects of Properties in Mathematics: Quantitative Characterizations
by Adrian I. Ban & Sorin G. Gal

Series on Concrete and Applicable Mathematics – Vol. 6

TOPICS ON STABILITY AND PERIODICITY IN ABSTRACT DIFFERENTIAL EQUATIONS

James H Liu
James Madison University, USA

Gaston M N'Guérékata
Morgan State University, USA

Nguyen Van Minh
University of West Georgia, USA

World Scientific

NEW JERSEY • LONDON • SINGAPORE • BEIJING • SHANGHAI • HONG KONG • TAIPEI • CHENNAI

Published by

World Scientific Publishing Co. Pte. Ltd.

5 Toh Tuck Link, Singapore 596224

USA office: 27 Warren Street, Suite 401-402, Hackensack, NJ 07601

UK office: 57 Shelton Street, Covent Garden, London WC2H 9HE

British Library Cataloguing-in-Publication Data
A catalogue record for this book is available from the British Library.

TOPICS ON STABILITY AND PERIODICITY IN ABSTRACT DIFFERENTIAL EQUATIONS
Series on Concrete and Applicable Mathematics — Vol. 6

ISBN-13 978-981-281-823-2
ISBN-10 981-281-823-5

Printed in Singapore.

Preface

Asymptotic behavior of evolution equations is a well-studied area in the theory of abstract differential equations with various methods of studies. It is natural to use the well-known ideas and techniques in the finite dimensional case as much as possible to deal with the problems in the infinite dimensional case. Having this in mind, in this book we will make an attempt to gather systematically certain recent results on several central topics of the asymptotic behavior of differential equations in Banach spaces. We will discuss the conditions for the stability, dichotomy and harmonic oscillation of solutions of evolution equations. The results and methods of approach will be presented in a manner that allows the reader, who is familiar with the techniques in the finite dimensional case, to easily understand them. Some parts of the book are actually lecture notes we have taught to graduate students over the past years.

We outline briefly the contents of our book. In Chapter 1 we recall several basic facts from semigroup theory, spectral theory of functions that will be used throughout the book. Chapter 2 is devoted to some classical topics including stability and dichotomy of linear homogeneous equations. In Chapter 3 we present some new methods of studying the harmonic oscillation in inhomogeneous linear equations. Chapter 4 is devoted to the topic of almost automorphy of solutions, that has recently regained interest in the mathematical literature. Existence of almost automorphic solutions to some linear and semilinear abstract differential equations is studied. We discuss the Massera type conditions for the existence of periodic solutions to periodic nonlinear equations in Chapter 5. At the end of each chapter we give a guide for further reading and comments on the results as well as the methods of study discussed in the chapter. We finally collect some of the required tools from functional analysis and operator theory in the appendices.

We wish to thank our colleagues and students for their encouragement and patience during the last years.

James H. Liu, Gaston M. N'guerekata, Nguyen Van Minh

Contents

Chapter 1

Banach Spaces, C_0-Semigroups of Linear Operators and Almost Periodicity of Functions

1.1 Banach Spaces and Linear Operators

1.1.1 *Banach Spaces*

The notion of Banach spaces will be used throughout this book. A normed space is a linear space $\mathbb{X}$, endowed with a norm, frequently denoted by $\|\cdot\|$, that is a function from $\mathbb{X}$ to the set of all real numbers, denoted by $\mathbb{R}$, such that

(1) $\|x\| \geq 0, \|x\| = 0$ if and only if $x = 0$;
(2) $\|\lambda x\| = |\lambda|\|x\|,\ \forall\lambda \in \mathbb{C}$ or $\mathbb{R}, x \in \mathbb{X}$;
(3) $\|x + y\| \leq \|x\| + \|y\|,\ \forall\ x, y \in \mathbb{X}$.

A normed space $\mathbb{X}$ is called *Banach space* if it is complete, i.e., every Cauchy sequence in $\mathbb{X}$ is convergent.

Example 1.1. Let $BC(\mathbb{R}, \mathbb{X})$ be the linear space of all bounded continuous $\mathbb{X}$-valued functions on $\mathbb{R}$ with sup-norm

$$\|f\| := \sup_{t\in\mathbb{R}} \|f(t)\|,\ \forall f \in BC(\mathbb{R}, \mathbb{X}). \tag{1.1}$$

Then $BC(\mathbb{R}, \mathbb{X})$ is a Banach space.

Similarly, the following spaces are Banach spaces if endowed with norm (1.1).

$$BUC(\mathbb{R}, \mathbb{X}) := \{f \in BC(\mathbb{R}, \mathbb{X}) : f \text{ is uniformly continuous}\} \tag{1.2}$$

$$\mathcal{P}_\omega := \{f \in BC(\mathbb{R}, \mathbb{X}) : f \text{ is periodic with period } \omega\} \tag{1.3}$$

However, the function space

$$C^1(\mathbb{R}, \mathbb{X}) := \{f \in BC(\mathbb{R}, \mathbb{X}) : f' \text{ exists and } f' \in BC(\mathbb{R}, \mathbb{X})\}$$

is not a Banach space. In fact, it is easy to choose a sequence of differentiable functions $\{f_n\}$ with $f_n' \in BC(\mathbb{R}, \mathbb{X})$ such that it is a Cauchy sequence but it does not converge to a differentiable function.

Example 1.2. Let Ω be the unit open ball of $\mathbb{R}^n$, i.e., $\Omega = \{x \in \mathbb{R}^n : \|x\| < 1\}$. We denote by $C^m(\overline{\Omega})$ the set of all m times continuously differentiable functions in Ω with the derivatives up to the order m bounded and continuously extendable up to the boundary $\{x \in \mathbb{R}^n : \|x\| = 1\}$. Then $C^m(\overline{\Omega})$ is a Banach space with the following norm

$$\|f\|_{C^m(\overline{\Omega})} := \sum_{|\alpha| \le m} \sup_{x \in \Omega} \|D^\alpha f(x)\|. \tag{1.4}$$

To conclude this subsection we consider the following Banach spaces

Example 1.3. For any interval $I = [a, b], a < b \in \mathbb{R}$ and $\alpha \in (0, 1)$ let us denote

$$C^\alpha(I, \mathbb{X}) := \{f \in BC(I, \mathbb{X}) : \sup_{t,s \in I, s<t} \frac{\|f(t) - f(s)\|}{(t-s)^\alpha} < \infty\}.$$

Then $C^\alpha(I, \mathbb{X})$ is a Banach space with the norm

$$\|f\|_{C^\alpha(I,\mathbb{X})} := \sup_{t \in I} \|f(t)\| + \sup_{t,s \in I, s<t} \frac{\|f(t) - f(s)\|}{(t-s)^\alpha}.$$

These Banach spaces are called *the Banach spaces of Hölder continuous functions.*

In general if $dim\mathbb{X} < \infty$, then $\mathbb{X}$ is a Banach space with any norm.

Exercise 1. Show that $\mathbb{X}$ with norm $\|\cdot\|$ is a finite-dimensional Banach space if and only if the unit ball $\{x \in \mathbb{X} : \|x\| \le 1\}$ is compact.

1.1.2 *Linear Operators*

Definition 1.1. Let $\mathbb{X}$ be a Banach space. Then a mapping A from $D(A) \subset \mathbb{X}$ to $\mathbb{X}$ is said to be a *linear operator* if $D(A)$ is a linear subspace of $\mathbb{X}$ and A is linear. In this case $D(A)$ is called *domain* of A and the range of this operator will be denoted by $R(A)$.

Remark 1.1. In the definition of a linear operator the domain is necessarily a linear space. In general, it is a dense subspace of $\mathbb{X}$ but not the whole space $\mathbb{X}$.

Example 1.4. Let M be an $n \times n$ matrix with real entries. Then it defines a linear operator from $\mathbb{R}^n$ into itself by the rule $x \mapsto Mx$, where x is a columm of n rows, an element of $\mathbb{R}^n$. In this example, denoting by $\mathcal{M}$ the corresponding linear operator, we have that $D(\mathcal{M}) = \mathbb{R}^n$.

Generally, if $dim\mathbb{X} < \infty$, then from the density of $D(A)$ in $\mathbb{X}$ follows that $D(A) = \mathbb{X}$. However, it is not the case for the following operators in infinite dimensional Banach spaces:

Example 1.5. Let A be the differential operator d/dt with $D(A)$ defined as follows:

$$D(A) = \{f \in BC(\mathbb{R}, \mathbb{X}) : df/dt \in BC(\mathbb{R}, \mathbb{X})\}.$$

Obviously, A is a linear operator and $D(A)$ is a subspace of $\mathbb{X}$ that is dense everywhere in $\mathbb{X}$, but is not $\mathbb{X}$.

Example 1.6. Let B be the differential operator d/dt with $D(B)$ defined as follows:

$$D(B) := \{g \in BC(\mathbb{R}, \mathbb{X}) : dg/dt \in BC(\mathbb{R}, \mathbb{X}), dg(0)/dt = 1\}.$$

It is not difficult to see that B is not a linear subspace of $BC(\mathbb{R}, \mathbb{X})$ as the domain $D(B)$ is not linear subspace. In fact, we can see that 0 is not in $D(B)$.

1.1.3 *Spectral Theory of Linear (Closed) Operators*

First, we introduce the notion of bounded linear operators on a Banach space $\mathbb{X}$, and then extend our consideration to more general classes of linear operators, for instance, closed operators.

Definition 1.2. Let A be a linear operator on a Banach space $\mathbb{X}$ with $D(A) = \mathbb{X}$. Then it is said to be a bounded linear operator on $\mathbb{X}$ if there exists a positive constant c such that

$$\|Ax\| \leq c\|x\|, \ \forall x \in \mathbb{X}. \tag{1.5}$$

Hence, if A is a bounded linear operator, then it is continuous.

Exercise 2. Show the converse of the above assertion.

In view of this exercise, the notion of bounded linear operators is nothing but that of continuous linear operators.

Let A be a bounded linear operator. Then the following nonnegative number

$$\|A\| := inf\{c \in \mathbb{R} : \|Ax\| \leq c\|x\|, \ \forall x \in \mathbb{X}\} \tag{1.6}$$

is called *the norm of A.*

Exercise 3. Let $L(\mathbb{X})$ denote the set of all bounded linear operators on the given Banach space $\mathbb{X}$. Then $L(\mathbb{X})$ endowed with the norm (1.6) is again a Banach space.

As we have seen in the above example, there are linear operators that are not bounded. Among the class of unbounded linear operators the class of closed operators is particularly important. There are two reasons, to our view, for this importance. The first one is that we encounter the operators of this class everywhere in problems involving partial differential equations, functional differential equations, integro-differential equations, etc. The second one is that the requirement on the closedness is indeed not too much. Every linear operator with nonempty resolvent set is closed, as shown below. Now we give a precise definition of this class.

Definition 1.3. A linear operator A from $D(A) \subset \mathbb{X}$ to $\mathbb{X}$ is said to be closed if its graph, i.e., the set $\{(x, Ax) \in \mathbb{X} \times \mathbb{X}, \forall x \in D(A)\}$ is closed.

If a linear operator A is not closed, one may expect that there is an extension so that the extension of it is closed. In this case, we say that the linear operator A is closable. The smallest extension is called the *closure* of A. We are ready to define the notion of spectrum of a closed linear operator A. Let X be a given complex Banach space.

Definition 1.4. We call the set

$$\rho(A) := \{\lambda \in \mathbb{C} : \lambda - A : D(A) \to \mathbb{X} \text{ is bijective}\}$$

the resolvent set and its complement $\sigma(A) := \mathbb{C} \backslash \rho(A)$ the spectrum of A. For $\lambda \in \rho(A)$, the inverse

$$R(\lambda, A) := (\lambda - A)^{-1}$$

is, by the closed graph theorem, a bounded linear operator on $\mathbb{X}$ and will be called the resolvent of A in the point λ.

Remark 1.2. Definition (1.4) can be extended to general linear operators, which are not necessarily closed, if we require that the map $A : D(A) \subset \mathbb{X} \to \mathbb{X}$ has bounded inverse.

Exercise 4. Show that if a linear operator $A : D(A) \subset \mathbb{X} \to \mathbb{X}$ has nonempty resolvent set, then it is closed.

Example 1.7. Let A be any linear operator of the finite dimensional complex Banach space $\mathbb{C}^n$. Then $\sigma(A)$ is exactly the set of all eigenvalues of A, i.e. the set of all $\lambda \in \mathbb{C}$ such that there is a nonzero vector x of $\mathbb{C}^n$ with $Ax = \lambda x$.

Any bounded linear operator A has nonempty spectrum. Its spectrum is contained in the disk of radius

$$r(A) := \lim_{n\to\infty} \left(\|A\|^n\right)^{1/n}. \tag{1.7}$$

This number is called *spectral radius* of the bounded operator A.

In contrast to the finite dimensional case, in general, a bounded linear spectrum may have no eigenvalue, for instance,

Example 1.8. Let c_0 be the Banach space of numerical two-sided sequences which converge to zero with sup-norm, i.e.,

$$c_0 := \{\{x_n\}_{n=0}^{\infty}, x_n \in \mathbb{R}, \lim_{n\to\infty} x_n = 0\}.$$

Then we define the translation $T : c_0 \to c_0$ which maps every sequence $\{x_n\}$ to the sequence $\{y_n\}$ such that $y_n = x_{n-1}$, $\forall n > 1, y_1 = 0$. It is seen that

$$\bigcap_{n=1}^{\infty} T^n c_0 = \{0\}.$$

So, if there is an eigenvalue λ, then there is an invariant nontrivial subspace of T. This is impossible.

1.1.3.1 *Several Properties of Resolvents*

An important property of $\rho(A)$ is that it is an open subset of the complex plane. The map $\rho(A) \ni \lambda \to R(\lambda, A) \in L(\mathbb{X})$ (*resolvent map*) is analytic in the open subset $\rho(A)$. One can show that any closed subset of the complex plane can serve as a spectrum of a closed linear operator.

Theorem 1.1. *For a closed linear operator $A : D(A) \subset \mathbb{X} \to \mathbb{X}$, the following properties hold true:*

(1) The resolvent set $\rho(A)$ is open in $\mathbb{C}$, and for $\mu \in \rho(A)$ one has

$$R(\lambda, A) = \sum_{n=0}^{\infty} (\mu - \lambda)^n R(\mu, A)^{n+1} \tag{1.8}$$

for all $\lambda \in \mathbb{C}$ such that $|\mu - \lambda| < 1/\|R(\mu, A)\|$.

(2) *The resolvent map* $\lambda \mapsto R(\lambda, A)$ *is locally analytic with*

$$\frac{d^n}{d\lambda^n} R(\lambda, A) = (-1)^n n! R(\lambda, A)^{n+1}, \ \forall n \in \mathbf{N}.$$

(3) *Let* $\lambda_n \in \rho(A)$ *with limit* $\lim_{n\to\infty} \lambda_n = \lambda_0$. *Then* $\lambda_0 \in \sigma(A)$ *if and only if*

$$\lim_{n\to\infty} \|R(\lambda_n, A)\| = \infty.$$

Proof. For the proof see e.g. Chap. IV in [Engel and Nagel (30)]. □

The following is concerned with another elementary property of resolvents:

Theorem 1.2. *Let A be a closed linear operator on a Banach space X. Then for all* $\lambda \in \rho(A)$ *we have*

$$\|R(\lambda, A)\| \geq \frac{1}{dist\,(\lambda, \sigma(A))}.$$

Proof. This will be an immediate consequence of the fact that $\|R(\lambda, A)\| \geq r(R(\lambda, A))$ once we prove that

$$\sigma(R(\lambda, A)) = \frac{1}{\lambda - \sigma(A)}.$$

But it is trivial to check that $(\lambda - \mu)(\lambda - A)R(\mu, A)$ is a two-sided inverse for $(\lambda - \mu)^{-1} - R(\lambda, A)$ whenever $\mu \in \rho(A)$, which proves the inclusion $\subset$. Similarly, $(\lambda - \mu)^{-1} R(\lambda, A)((\lambda - \mu)^{-1} - R(\lambda, A))^{-1}$ is a two-sided inverse for $\mu - A$ whenever $(\lambda - \mu)^{-1} \in \rho(R(\lambda, A))$, which proves the inclusion $\supset$. □

Let A be a closed linear operator. Then we introduce some finer notions of spectrum for A.

Definition 1.5.

(1) Each $\lambda \in \sigma(A)$ such that $\lambda - A$ is not injective, is called eigenvalue of A, and each nonzero vector $x \in D(A)$ such that $(\lambda - A)x = 0$ is called eigenvector corresponding to λ. The subset of $\sigma(A)$ consisting of all eigenvalues of A, is denoted by $P\sigma(A)$ and is called the point spectrum of A.
(2) We call approximate eigenvalue each $\lambda \in \sigma(A)$ such that there is a sequence $\{x_n\} \subset D(A)$ (called approximate eigenvector) satisfying $\|x_n\| = 1$ and $\lim_{n\to\infty} \|Ax_n - \lambda x_n\| = 0$.

1.2 Strongly Continuous Semigroups of Operators

In this section we collect some well-known facts from the theory of strongly continuous semigroups of operators on a Banach space for the reader's convenience. We will focus the reader's attention on several important classes of semigroups such as analytic and compact semigroups which will be discussed in the next chapters. Among the basic properties of strongly continuous semigroups we will emphasize on the spectral mapping theorem. Since the materials of this section as well as of this chapter in the whole can be found in any standard book covering the area, here we aim at freshening up the reader's memory rather than giving a logically self contained account of the theory.

1.2.1 *Definition and Basic Properties*

Definition 1.6. A family $(T(t))_{t\geq 0}$ of bounded linear operators acting on a Banach space $\mathbb{X}$ is called a C_0-*semigroup* if the following three properties are satisfied:

(1) $T(0) = I$, the identity operator on $\mathbb{X}$;
(2) $T(t)T(s) = T(t+s)$ for all $t, s \geq 0$;
(3) $\lim_{t\to 0^+} \|T(t)x - x\| = 0$ for all $x \in \mathbb{X}$.

The *infinitesimal generator* of $(T(t))_{t\geq 0}$, or briefly the *generator*, is the linear operator A with domain $D(A)$ defined by

$$D(A) = \{x \in \mathbb{X} : \lim_{t\downarrow 0} \frac{1}{t}(T(t)x - x) \text{ exists}\},$$

$$Ax = \lim_{t\downarrow 0} \frac{1}{t}(T(t)x - x), \quad x \in D(A).$$

The generator is always a closed, densely defined operator. A strongly continuous semigroup of bounded linear operators on $\mathbb{X}$ will be called C_0-*semigroup*.

We now consider several examples of strongly continuous semigroups.

Example 1.9. Let A be a bounded linear operator on a Banach space $\mathbb{X}$. Then $(e^{tA}))_{t\geq 0}$, defined by the formula

$$e^{tA} := \sum_{k=0}^{\infty} \frac{(tA)^k}{k!}, \tag{1.9}$$

is a strongly continuous semigroup of bounded linear operators on the Banach space $\mathbb{X}$. Moreover, its generator is the operator A with $D(A) = \mathbb{X}$.

Proof. First, it may be noted that the formula (1.9) is well defined. In fact, since A is bounded

$$\sum_{k=0}^{\infty} \frac{\|(tA)^k\|}{k!} \leq e^{|t|\|A\|}, \ \forall t.$$

Hence, the series is absolutely convergent. To check that this family is indeed a semigroup we will prove the following

$$e^{(t+s)A} = e^{tA}e^{sA}, \ \forall s, t \in \mathbb{R}.$$

From the absolute convergence of the above series it follows that the product of two series $\sum_{k=0}^{\infty} \frac{(tA)^k}{k!}$, and $e^{sA} := \sum_{n=0}^{\infty} \frac{(sA)^n}{n!}$ is absolutely convergent, i.e.,

$$\sum_{k=0}^{\infty} \frac{(tA)^k}{k!} \times \sum_{n=0}^{\infty} \frac{(sA)^n}{n!}$$

is convergent. Moreover, it does not depend on the way of summation, in particular,

$$\begin{aligned}
e^{tA}e^{sA} &= \sum_{k=0}^{\infty} \frac{(tA)^k}{k!} \times \sum_{n=0}^{\infty} \frac{(sA)^n}{n!} \\
&= \sum_{m=0}^{\infty} \sum_{k+n=m} \frac{(tA)^k}{k!} \frac{(sA)^n}{n!} &(1.10)\\
&= \sum_{m=0}^{\infty} \frac{(tA)^m}{m!} &(1.11)\\
&= e^{(t+s)A}. &(1.12)
\end{aligned}$$

We now show that this semigroup is strongly continuous. By definition, we have to show that

$$\lim_{t\to 0^+} e^{tA}x = x, \ \forall x \in \mathbb{X}.$$

By (1.9), $\forall x \in \mathbb{X}$

$$\begin{aligned}
\|e^{tA}x - x\| &\leq \| \sum_{k=1}^{\infty} \frac{(tAx)^k x}{k!} \| \\
&\leq |t|\|A\|\|x\| \sum_{k=0}^{\infty} \frac{\|(tA)^k\|}{(k+1)!} \\
&\leq |t|\|A\|\|x\| \sum_{k=0}^{\infty} \frac{\|(tAx)^k\|}{(k)!} \\
&= |t|\|A\|\|x\| e^{|t|\|A\|}.
\end{aligned}$$

Hence

$$\lim_{t\to 0^+} \|e^{tA}x - x\| = 0.$$

Now we show that A is the generator of this semigroup with $D(A) = \mathbb{X}$, i.e., we have to show that for all $x \in \mathbb{X}$, the following holds true

$$\lim_{t\to 0^+} \frac{e^{tA}x - x}{t} = Ax. \tag{1.13}$$

In fact, by (1.9), $\forall x \in \mathbb{X}$

$$\|\frac{e^{tA}x - x}{t} - Ax\| \le |t|e^{|t|\|A\|}\|A\|\|x\|. \tag{1.14}$$

Hence (1.13) holds true. □

Example 1.10. Let $(S(t))_{t\ge 0}$ be the translation semigroup on $BUC(\mathbb{R}, \mathbb{X})$, where $\mathbb{X}$ is a Banach space, i.e.,

$$S(t)f(s) := f(t+s), \ \forall t \ge 0, s \in \mathbb{R}, f \in BUC(\mathbb{R}, \mathbb{X}).$$

It is easy to check that $(S(t))_{t\ge 0}$ is a semigroup of bounded linear operators on $BUC(\mathbb{R}, \mathbb{X})$. From the uniform continuity of every function in $BUC(\mathbb{R}, \mathbb{X})$ it follows that this translation semigroup is strongly continuous.

As an example of a non strongly continuous semigroup we can consider the translation semigroup $(S(t))_{t\ge 0}$ in $BC(\mathbb{R}, \mathbb{X})$. The non-strong continuity follows from the fact that there exists a bounded function which is not uniformly continuous. Indeed, we can take the following example

Example 1.11. The function $f(t) = \sin t^2$, $t \in \mathbb{R}$, is a continuous and bounded function that is not uniformly continuous.

In fact, we can choose a sequence $t_{2n} = \sqrt{2n\pi + \pi/2}$, $t_{2n+1} = \sqrt{(2n+1)\pi}$. Obviously, $f(t-2n) = 1$ and $f(t_{2n+1}) = 0$ while $|t_{2n} - t_{2n+1}| \to 0$ as $n \to \infty$.

Theorem 1.3. *Let $(T(t))_{t\ge 0}$ be a C_0-semigroup. Then there exist constants $\omega \ge 0$ and $M \ge 1$ such that*

$$\|T(t)\| \le Me^{\omega t}, \ \forall t \ge 0.$$

Proof. For the proof see e.g. p. 4 in [Pazy (90)]. □

Corollary 1.1. *If $(T(t))_{t\ge 0}$ is a C_0-semigroup, then the mapping $(x,t) \mapsto T(t)x$ is a continuous function from $\mathbb{X} \times \mathbb{R}^+ \to \mathbb{X}$.*

Proof. For any $x, y \in \mathbb{X}$ and $t \leq s \in \mathbb{R}^+ := [0, \infty)$,

$$\begin{aligned} \|T(t)x - T(s)y\| &\leq \|T(t)x - T(s)x\| + \|T(s)x - T(s)y\| \\ &\leq Me^{\omega s}\|x - y\| + \|T(t)\|\|T(s-t)x - x\| \\ &\leq Me^{\omega s}\|x - y\| + Me^{\omega t}\|T(s-t)x - x\|. \end{aligned} \tag{1.15}$$

Hence, for fixed x, t ($t \leq s$) if $(y, s) \to (x, t)$ then $\|T(t)x - T(s)y\| \to 0$. Similarly, for $s \leq t$

$$\begin{aligned} \|T(t)x - T(s)y\| &\leq \|T(t)x - T(s)x\| + \|T(s)x - T(s)y\| \\ &\leq Me^{\omega s}\|x - y\| + \|T(s)\|\|T(t-s)x - x\| \\ &\leq Me^{\omega s}\|x - y\| + Me^{\omega s}\|T(t-s)x - x\|. \end{aligned} \tag{1.16}$$

Hence, if $(y, s) \to (x, t)$ then $\|T(t)x - T(s)y\| \to 0$. □

Other basic properties of a C_0-semigroup and its generator are listed in the following:

Theorem 1.4. *Let A be the generator of a C_0-semigroup $(T(t))_{t\geq 0}$ on $\mathbb{X}$. Then*

(1) For $x \in \mathbb{X}$,

$$\lim_{h\to 0} \frac{1}{h} \int_t^{t+h} T(s)xds = T(t)x.$$

(2) For $x \in \mathbb{X}$, $\int_0^t T(s)xds \in D(A)$ and

$$A\left(\int_0^t T(s)xds\right) = T(t)x - x.$$

(3) For $x \in D(A), T(t)x \in D(A)$ and

$$\frac{d}{dt}T(t)x = AT(t)x = T(t)Ax.$$

(4) For $x \in D(A)$,

$$T(t)x - T(s)x = \int_s^t T(\tau)Axd\tau = \int_s^t AT(\tau)xd\tau.$$

Proof. For the proof see e.g. p. 5 in [Pazy (90)]. □

We continue with some useful facts about semigroups that will be used throughout this book. The first of these is the *Hille-Yosida theorem* that characterizes the generators of C_0-semigroups among the class of all linear operators.

Theorem 1.5. *Let A be a linear operator on a Banach space $\mathbf{X}$, and let $\omega \in \mathbb{R}$ and $M \geq 1$ be constants. Then the following assertions are equivalent:*

(1) *A is the generator of a C_0-semigroup $(T(t))_{t\geq 0}$ satisfying $\|T(t)\| \leq Me^{\omega t}$ for all $t \geq 0$;*
(2) *A is closed, densely defined, the half line (ω, ∞) is contained in the resolvent set $\rho(A)$ of A, and we have the estimates*

$$\|R(\lambda, A)^n\| \leq \frac{M}{(\lambda - \omega)^n}, \qquad \forall \lambda > \omega, \quad n = 1, 2, ... \tag{1.17}$$

Here, $R(\lambda, A) := (\lambda - A)^{-1}$ denotes the resolvent of A at λ. If one of the equivalent assertions of the theorem holds, then actually $\{Re\lambda > \omega\} \subset \rho(A)$ and

$$\|R(\lambda, A)^n\| \leq \frac{M}{(Re\lambda - \omega)^n}, \qquad \forall Re\lambda > \omega, \quad n = 1, 2, ... \tag{1.18}$$

Moreover, for $Re\lambda > \omega$ the resolvent is given explicitly by

$$R(\lambda, A)x = \int_0^\infty e^{-\lambda t} T(t)x\, dt, \qquad \forall x \in \mathbb{X}. \tag{1.19}$$

We shall mostly need the implication (i)⇒(ii), which is the easy part of the theorem. In fact, one checks directly from the definitions that

$$R_\lambda x := \int_0^\infty e^{-\lambda t} T(t)x\, dt$$

defines a two-sided inverse for $\lambda - A$. The estimate (1.18) and the identity (1.19) follow trivially from this.

A useful consequence of (1.17) is that

$$\lim_{\lambda \to \infty} \|\lambda R(\lambda, A)x - x\| = 0, \quad \forall x \in X. \tag{1.20}$$

This is proved as follows. Fix $x \in D(A)$ and $\mu \in \rho(A)$, and let $y \in X$ be such that $x = R(\mu, A)y$. By (1.17) we have $\|R(\lambda, A)\| = O(\lambda^{-1})$ as $\lambda \to \infty$. Therefore, the *resolvent identity*

$$R(\lambda, A) - R(\mu, A) = (\mu - \lambda)R(\lambda, A)R(\mu, A) \tag{1.21}$$

implies that

$$\lim_{\lambda \to \infty} \|\lambda R(\lambda, A)x - x\| = \lim_{\lambda \to \infty} \|R(\lambda, A)(\mu R(\mu, A)y - y)\| = 0.$$

This proves (1.20) for elements $x \in D(A)$. Since $D(A)$ is dense in X and the operators $\lambda R(\lambda, A)$ are uniformly bounded as $\lambda \to \infty$ by (1.17), (1.20) holds for all $x \in \mathbb{X}$.

1.2.2 Compact Semigroups and Analytic Strongly Continuous Semigroups

Definition 1.7. A C_0-semigroup $(T(t))_{t\geq 0}$ is called *compact* for $t > t_0$ if for every $t > t_0$, $T(t)$ is a compact operator. $(T(t))_{t\geq 0}$ is called *compact* if it is compact for each $t > 0$.

If a C_0-semigroup $(T(t))_{t\geq 0}$ is compact for $t > t_0$, then it is continuous in the uniform operator topology for $t > t_0$.

Theorem 1.6. *Let A be the generator of a C_0-semigroup $(T(t))_{t\geq 0}$. Then $(T(t))_{t\geq 0}$ is a compact semigroup if and only if $T(t)$ is continuous in the uniform operator topology for $t > 0$ and $R(\lambda; A)$ is compact for $\lambda \in \rho(A)$.*

Proof. For the proof see e.g. p. 49 in [Pazy (90)]. □

In this book we distinguish the notion of analytic C_0-semigroups from that of analytic semigroups in general. To this end we recall several notions. Let A be a linear operator $D(A) \subset \mathbb{X} \to \mathbb{X}$ with *not necessarily dense domain.*

Definition 1.8. A is said to be *sectorial* if there are constants $\omega \in \mathbb{R}, \theta \in (\pi/2, \pi), M > 0$ such that the following conditions are satisfied:

$$\begin{cases} i) \ \ \rho(A) \supset S_{\theta,\omega} = \{\lambda \in \mathbb{C} : \lambda \neq \omega, |arg(\lambda - \omega)| < \theta\}, \\ \\ ii) \ \ \|R(\lambda, A)\| \leq M/|\lambda - \omega| \ \forall \lambda \in S_{\theta,\omega}. \end{cases}$$

If we assume in addtion that $\rho(A) \neq \oslash$, then A is closed. Thus, $D(A)$, endowed with the graph norm

$$\|x\|_{D(A)} := \|x\| + \|Ax\|,$$

is a Banach space. For a sectorial operator A, from the definition, we can define a linear bounded operator e^{tA} by means of the Dunford integral

$$e^{tA} := \frac{1}{2\pi i} \int_{\omega+\gamma_{r,\eta}} e^{t\lambda} R(\lambda, A) d\lambda, t > 0, \tag{1.22}$$

where $r > 0, \eta \in (\pi/2, \theta)$ and $\gamma_{r,\eta}$ is the curve

$$\{\lambda \in \mathbb{C} : |arg\lambda| = \eta, |\lambda| \geq r\|\} \cup \{\lambda \in \mathbb{C} : |arg\lambda| \leq \eta, |\lambda| = r\},$$

oriented counterclockwise. In addition, set $e^{0A}x = x, \ \forall x \in \mathbb{X}$.

Theorem 1.7. *Under the above notation, for a sectorial operator A the following assertions hold true:*

(1) $e^{tA}x \in D(A^k)$ *for every* $t > 0, x \in \mathbb{X}, k \in \mathbf{N}$. *If* $x \in D(A^k)$, *then*

$$A^k e^{tA}x = e^{tA}A^k x, \ \forall t \geq 0;$$

(2) $e^{tA}e^{sA} = e^{(t+s)A}, \ \forall t, s \geq 0;$

(3) There are positive constants $M_0, M_1, M_2, ...,$ *such that*

$$\begin{cases} (a) \ \ \|e^{tA}\| \leq M_0 e^{\omega t}, \ t \geq 0, \\ \\ (b) \ \ \|t^k (A - \omega I)^k e^{tA}\| \leq M_k e^{\omega t}, \ t \geq 0, \end{cases}$$

where ω *is determined from Definition 1.8. In particular, for every* $\varepsilon > 0$ *and* $k \in \mathbf{N}$ *there is* $C_{k,\varepsilon}$ *such that*

$$\|t^k A^k e^{tA}\| \leq C_{k,\varepsilon} e^{(\omega+\varepsilon)t}, \ t > 0;$$

(4) The function $t \mapsto e^{tA}$ *belongs to* $C^\infty((0,+\infty), L(\mathbb{X}))$, *and*

$$\frac{d^k}{dt^k} e^{tA} = A^k e^{tA}, \ t > 0,$$

moreover it has an analytic extension in the sector

$$S = \{\lambda \in \mathbb{C} : |arg\lambda| < \theta - \pi/2\}.$$

Proof. For the proof see pp. 35-37 in [Lunardi (65)]. □

Definition 1.9. For every sectorial operator A the semigroup $(e^{tA})_{t \geq 0}$ defined in Theorem 1.7 is called *the analytic semigroup* generated by A in $\mathbb{X}$. An analytic semigroup is said to be an *analytic strongly continuous semigroup* if in addition, it is strongly continuous.

There are analytic semigroups which are not strongly continuous, for instance, the analytic semigroups generated by nondensely defined sectorial operators. From the definition of sectorial operators it is obvious that for a sectorial operator A the intersection of the spectrum $\sigma(A)$ with the imaginary axis is bounded. In this book, if otherwise stated, by "analytic semigroups" we mean analytic semigroups that are strongly continuous. We use this convention because most of the results presented here are concerned with C_0-semigroups.

1.2.3 *Spectral Mapping Theorems*

If A is a bounded linear operator on a Banach space $\mathbb{X}$, then by the Dunford Theorem [Dunford and Schwartz (29)] $\sigma(exp(tA)) = exp(t\sigma(A))$, $\forall t \geq 0$. It is natural to expect this relation holds for any C_0-semigroups on a Banach space. However, this is not true in general as shown by the following counterexample (p. 44 in [Pazy (90)])

Example 1.12.

Let $\mathbb{X}$ be the Banach space of all continuous functions on the interval $[0,1]$ which are equal to zero at $x = 1$ with the supremum norm. Define

$$(T(t)f)(x) = \begin{cases} f(x+t), \text{ if } x+t \leq 1 \\ 0, \text{ if } x+t > 1. \end{cases}$$

$(T(t))_{t\geq 0}$ is obviously a C_0-semigroup of contraction on $\mathbb{X}$. Its generator A is given by

$$D(A) = \{f : f \in C^1([0,1]) \cap \mathbb{X}, f' \in \mathbb{X}\}$$

and

$$Af = f', \quad \text{for } f \in D(A).$$

For every $\lambda \in \mathbb{C}$ and $g \in \mathbb{X}$ the equation $\lambda f - f' = g$ has a unique solution $f \in \mathbb{X}$ given by

$$f(t) = \int_t^1 e^{\lambda(t-s)} g(s) ds.$$

Therefore, $\sigma(A) = \emptyset$. On the other hand, since for every $t \geq 0$, $T(t)$ is a bounded operator, $\sigma(T(t)) \neq 0$, so the relation $\sigma(T(t)) = exp(t\sigma(A))$ does not hold for any $t \geq 0$.

In this section we prove the following Spectral Inclusion Theorem for C_0-semigroups:

Theorem 1.8. *Let $(T(t))_{t\geq 0}$ be a C_0-semigroup on a Banach space X with generator A. Then we have the spectral inclusion relation*

$$\sigma(T(t)) \supset e^{t\sigma(A)}, \quad \forall t \geq 0.$$

Proof. By Theorem 1.4 for the semigroup $(T_\lambda(t))_{t\geq 0} := \{e^{-\lambda t}T(t)\}_{t\geq 0}$ generated by $A - \lambda$, for all $\lambda \in \mathbb{C}$ and $t \geq 0$

$$(\lambda - A)\int_0^t e^{\lambda(t-s)} T(s)x \, ds = (e^{\lambda t} - T(t))x, \quad \forall x \in X,$$

and

$$\int_0^t e^{\lambda(t-s)}T(s)(\lambda - A)x\,ds = (e^{\lambda t} - T(t))x, \quad \forall x \in D(A). \qquad (2.1.1)$$

Suppose $e^{\lambda t} \in \rho(T(t))$ for some $\lambda \in \mathbb{C}$ and $t \geq 0$, and denote the inverse of $e^{\lambda t} - T(t)$ by $Q_{\lambda,t}$. Since $Q_{\lambda,t}$ commutes with $T(t)$ and hence also with A, we have

$$(\lambda - A)\int_0^t e^{\lambda(t-s)}T(s)Q_{\lambda,t}x\,ds = x, \quad \forall x \in X,$$

and

$$\int_0^t e^{\lambda(t-s)}T(s)Q_{\lambda,t}(\lambda - A)x\,ds = x, \quad \forall x \in D(A).$$

This shows the boundedness of the operator B_λ defined by

$$B_\lambda x := \int_0^t e^{\lambda(t-s)}T(s)Q_{\lambda,t}x\,ds$$

is a two-sided inverse of $\lambda - A$. It follows that $\lambda \in \varrho(A)$. □

As shown by Example 1.12 the converse inclusion

$$e^{t\sigma(A)} \supset \sigma(T(t))\backslash\{0\}$$

in general fails. For certain parts of the spectrum, however, the Spectral Mapping Theorem holds true. To make it more clear we recall that for a given closed operator A on a Banach space $\mathbb{X}$ the *point spectrum* $\sigma_p(A)$ is the set of all $\lambda \in \sigma(A)$ for which there exists a non-zero vector $x \in D(A)$ such that $Ax = \lambda x$, or equivalently, for which the operator $\lambda - A$ is not injective; the *residual spectrum* $\sigma_r(A)$ is the set of all $\lambda \in \sigma(A)$ for which $\lambda - A$ does not have dense range; the *approximate point spectrum* $\sigma_a(A)$ is the set of all $\lambda \in \sigma(A)$ for which there exists a sequence (x_n) of norm one vectors in X, $x_n \in D(A)$ for all n, such that

$$\lim_{n\to\infty} \|Ax_n - \lambda x_n\| = 0.$$

Obviously, $\sigma_p(A) \subset \sigma_a(A)$.

Theorem 1.9. *Let $(T(t))_{t\geq 0}$ be a C_0-semigroup on a Banach space $\mathbb{X}$, with generator A. Then*

$$\sigma_p(T(t))\backslash\{0\} = e^{t\sigma_p(A)}, \quad \forall t \geq 0.$$

Proof. For the proof see e.g. p. 46 in [Pazy (90)]. □

Recall that a family of bounded linear operators $(T(t))_{t\in\mathbb{R}}$ is said to be a *strongly continuous group* if it satisfies

(1) $T(0) = I$,
(2) $T(t+s) = T(t)T(s),\ \forall t, s \in \mathbb{R}$,
(3) $\lim_{t\to 0} T(t)x = x,\ \forall x \in \mathbb{X}$.

Similarly as C_0-semigroups, the generator of a strongly continuous group $(T(t))_{t\in\mathbb{R}}$ is defined to be the operator

$$Ax := \lim_{t\to 0} \frac{T(t)x - x}{t},$$

with the domain $D(A)$ consisting of all elements $x \in \mathbb{X}$ such that the above limit exists.

In the next chapter we need the following lemma:

Lemma 1.1. *Let $(T(t))_{t\geq 0}$ be a uniformly bounded C_0-group on a Banach space $\mathbb{X} \neq \{0\}$, with generator A. Then $\sigma(A) \neq \emptyset$.*

For bounded strongly continuous groups of linear operators the following Weak Spectral Mapping Theorem holds:

Theorem 1.10. *Let $(T(t))_{t\in\mathbb{R}}$ be a bounded strongly continuous group, i.e., there exists a positive M such that $\|T(t)\| \leq M,\ \forall t \in \mathbb{R}$ with generator A. Then*

$$\sigma(T(t)) = \overline{e^{t\sigma(A)}},\ \forall t \in \mathbb{R}. \tag{1.23}$$

Proof. For the proof see e.g. [Nagel (73)] or Chapter 2 in [van Neerven (78)]. □

Example 1.13. Let $\mathcal{M}$ be a closed translation invariant subspace of the space of $\mathbb{X}$-valued bounded uniformly continuous functions on the real line $BUC(\mathbb{R}, \mathbb{X})$, i.e., $\mathcal{M}$ is closed and $S(t)\mathcal{M} \subset \mathcal{M},\ \forall t$, where $(S(t))_{t\in\mathbb{R}}$ is the translation group on $BUC(\mathbb{R}, \mathbb{X})$. Then

$$\sigma(S(t)|_{\mathcal{M}}) = \overline{e^{t\sigma(\mathcal{D}_{\mathcal{M}})}},\ \forall t \in \mathbb{R},$$

where $\mathcal{D}_{\mathcal{M}}$ is the generator of the restriction of translation group to $\mathcal{M}$.

In the next chapter we will consider situations similar to this example which arise in connection with invariant subspaces of so-called evolution semigroups.

1.2.4 *Commuting Operators*

Let A, B be two bounded linear operators on a given Banach space $\mathbb{X}$ which is assumed to be complex. The definition of commutativeness of these operators in this case is natural, i.e., the identity $AB = BA$ holds. In the general case, where the operators A and B are not necessarily everywhere defined, we have the following definition:

Definition 1.10. Let A and B be linear operators on a Banach space G with non-empty resolvent sets. We say that *A and B commute* if one of the following equivalent conditions hold:

(1) $R(\lambda, A)R(\mu, B) = R(\mu, B)R(\lambda, A)$ for some (all) $\lambda \in \rho(A), \mu \in \rho(B)$,
(2) $x \in D(A)$ implies $R(\mu, B)x \in D(A)$ and $AR(\mu, B)x = R(\mu, B)Ax$ for some (all) $\mu \in \rho(B)$.

Exercise 5. Show that if A, B are bounded linear operators which are commutative in the usual sense, i.e., $AB = BA$, then they are commuting operators in the sense of Definition 1.10.

Exercise 6. Show that if A is a bounded linear operator on a Banach space $\mathbb{X}$ and $(T(t))_{t\geq 0}$ is a strongly continuous semigroup on $\mathbb{X}$. Then, if A commutes with $T(t)$ for all $t \geq 0$, A must commute with the infinitesimal generator of the semigroup $(T(t))_{t\geq 0}$.

Let us consider the generator of the product of two commuting semigroups $(G(t))_{t\geq 0}, (H(t))_{t\geq 0}$, that is, the semigroup $(P(t))_{t\geq 0}$ defined by $P(t) = G(t) \cdot H(t)$. Generally, this semigroup may not be strongly continuous. Below, we assume that the product semigroup $(P(t))_{t\geq 0}$ is strongly continuous. Let us denote by $\mathcal{G}, \mathcal{H}, \mathcal{P}$ the generators of $(G(t))_{t\geq 0}$, $(H(t))_{t\geq 0}$, $(P(t))_{t\geq 0}$, respectively.

Exercise 7. Under the above assumptions and notations prove that

$$\mathcal{P} = \overline{\mathcal{G} + \mathcal{H}}.$$

Hint. First show that for every $t \geq 0$, $G(t)D(\mathcal{H}) \subset D(\mathcal{H})$ from which the following holds: $D(\mathcal{G}) \cap D(\mathcal{H}) \subset D(\mathcal{P})$. Next, for every $x \in \mathbb{X}, t > 0$ using the following

$$y(t) = \frac{1}{t^2} \int_0^t H(\xi) \int_0^t G(\eta) x d\xi d\eta \tag{1.24}$$

to prove that $D(\mathcal{G}) \cap D(\mathcal{H})$ is dense in $\mathbb{X}$. Using this fact, complete the proof by showing that $\overline{\mathcal{G} + \mathcal{H}} = \mathcal{P}$.

For $\theta \in (0, \pi), R > 0$ we denote $\Sigma(\theta, R) = \{z \in \mathbb{C} : |z| \geq R, |argz| \leq \theta\}$.

Definition 1.11. Let A and B be commuting operators. Then

(1) A is said to be of class $\Sigma(\theta + \pi/2, R)$ if there are positive constants θ, R such that $0 < \theta < \pi/2$, and

$$\Sigma(\theta + \pi/2, R) \subset \rho(A) \quad \text{and} \quad \sup_{\lambda \in \Sigma(\theta+\pi/2,R)} \|\lambda R(\lambda, A)\| < \infty, \tag{1.25}$$

(2) A and B are said to satisfy *condition P* if there are positive constants $\theta, \theta', R, \theta' < \theta$ such that A and B are of class $\Sigma(\theta + \pi/2, R), \Sigma(\pi/2 - \theta', R)$, respectively.

If A and B are commuting operators, $A + B$ is defined by $(A + B)x = Ax + Bx$ with domain $D(A + B) = D(A) \cap D(B)$.

We will use the following norm, defined by A on the space $\mathbf{X}$, $\|x\|_{\mathcal{T}_A} := \|R(\lambda, A)x\|$, where $\lambda \in \rho(A)$. It is seen that different $\lambda \in \rho(A)$ yields equivalent norms. We say that an operator C on $\mathbf{X}$ is A-closed if its graph is closed with respect to the topology induced by $\mathcal{T}_A$ on the product $\mathbf{X} \times \mathbf{X}$. It is easily seen that C is A-closable if $x_n \to 0, x_n \in D(C), Cx_n \to y$ with respect to $\mathcal{T}_A$ in $\mathbf{X}$ implies $y = 0$. In this case, A-closure of C is denoted by $\overline{C}^A$.

Theorem 1.11. *Assume that A and B commute. Then the following assertions hold:*

(1) If one of the operators is bounded, then

$$\sigma(A + B) \subset \sigma(A) + \sigma(B). \tag{1.26}$$

(2) If A and B satisfy condition P, then $A + B$ is A-closable, and

$$\sigma(\overline{(A + B)}^A) \subset \sigma(A) + \sigma(B). \tag{1.27}$$

In particular, if $D(A)$ is dense in $\mathbf{X}$, then $\overline{(A + B)}^A = \overline{A + B}$, where $\overline{A + B}$ denotes the usual closure of $A + B$.

Proof. For the proof we refer the reader to Theorems 7.2, 7.3 in [Arendt, Räbiger and Sourour (6)]. □

1.3 Spectral Theory and Almost Periodicity of Functions

1.3.1 *Introduction*

As is known, for a 2π-periodic continuous function $f : \mathbb{R} \to \mathbb{X}$ the Fourier exponents are defined to be the set:

$$\{\lambda \in \mathbf{Z} : \int_{-\pi}^{\pi} e^{-i\lambda\xi} f(\xi) d\xi \neq 0\}. \tag{1.28}$$

The notion of spectrum of a bounded function is a generalization of this notion of Fourier exponents. However, for any bounded function it is not expected that the integral in the above set is used, but instead of it another integral on the whole real line. This section will give a very short introduction to this spectral theory. We will take several examples to show how our abstract definition can be incorporated into simple cases in which the notions of Fourier, Bohr exponents are well known.

1.3.2 *Spectrum of a Bounded Function*

We denote by $\mathcal{F}$ the Fourier transform, i.e.

$$\hat{f}(s) := \int_{-\infty}^{+\infty} e^{-ist} f(t) dt \tag{1.29}$$

($s \in \mathbb{R}, f \in L^1(\mathbb{R})$). Then the *Beurling spectrum* of $u \in BUC(\mathbb{R}, \mathbb{X})$ is defined to be the following set

$$sp(u) := \{\xi \in \mathbb{R} : \forall \varepsilon > 0 \exists f \in L^1(\mathbb{R}), supp\hat{f} \subset (\xi - \varepsilon, \xi + \varepsilon), f * u \neq 0\} \tag{1.30}$$

where

$$f * u(s) := \int_{-\infty}^{+\infty} f(s-t) u(t) dt.$$

We consider the simplest case

Example 1.14. Let $f(t) = ae^{t\lambda t}$, where $\lambda \in \mathbb{R}, a \in \mathbb{X}$. Then $sp(f) = \lambda$.

Proof. As is well known, for any real $\mu \neq \lambda$ there exists a function $\phi \in L^1(\mathbb{R})$ such that the support of the Fourier transform of ϕ is contained in the interval $[\mu - \varepsilon, \mu + \varepsilon]$ for any ε such that $\lambda \notin [\mu - \varepsilon, \mu + \varepsilon]$. Hence, the convolution

$$
\begin{aligned}
\psi(s) &= \int_{-\infty}^{\infty} \phi(s-t)f(t)dt = \\
&= a\int_{-\infty}^{\infty} \phi(s-t)e^{i\lambda t}dt \\
&= -a\int_{-\infty}^{\infty} \phi(\xi)e^{i\lambda\xi+s}d\xi \\
&= -ae^{i\lambda s}\int_{-\infty}^{\infty} \phi(\xi)e^{i\lambda\xi} \\
&= -ae^{i\lambda s}\hat{\phi}(\lambda) \\
&= ae^{i\lambda s}\times 0 = 0. \qquad (1.31)
\end{aligned}
$$

□

This shows that $sp(f) \subset \{\lambda\}$. On the other hand, if in the above argument we take $\mu = \lambda$ and $\phi \in L^1(\mathbb{R})$ such that $\hat{\phi}(\lambda) = 1$, then $\psi(s) = ae^{i\lambda s} \neq 0$. This yields that $sp(f) = \{\lambda\}$.

As an immediate consequence of this example, the following holds true:

Example 1.15. Let

$$f(t) = \sum_{k=0}^{N} a_k e^{i\lambda_k t},$$

where $a_k \neq 0,\ \lambda_k \in \mathbb{R}, \forall k = 1, 2, ..., N$. Then $sp(f) = \{\lambda_1, ..., \lambda_N\}$.

Example 1.16. If $f(t)$ is a 2π-periodic function with Fourier series

$$\sum_{k\in \mathbf{Z}} a_k e^{2i\pi kt},$$

then

$$sp(f) = \{2\pi k : a_k \neq 0\}.$$

Proof. For every $\lambda \neq 2k_0\pi,\ k_0 \in \mathbf{Z}$ or $\lambda = 2k_0\pi$ at which $f_{k_0} = 0$, where f_n is the Fourier coefficients of f, and for every positive ε, let $\phi \in L^1(\mathbb{R})$ be a complex valued continuous function such that the support of its Fourier transform $supp\hat{\phi}(\xi) \subset [\lambda - \varepsilon, \lambda + \varepsilon]$. Put

$$u(t) = f * \phi(t) = \int_{-\infty}^{\infty} f(t-s)\phi(s)ds.$$

Since f is periodic, there is a sequence of trigonometric polynomials

$$P_n(t) = \sum_{k=1}^{N(n)} a_{n,k} e^{2ik\pi t}$$

that is convergent uniformly to f with respect to $t \in \mathbb{R}$ such that $\lim_{n\to\infty} a_{n,k} = f_n$. We have

$$\begin{aligned} u(t) = f * \phi(t) &= \lim_{n\to\infty} P_n * \phi(t) \\ &= \lim_{n\to\infty} \sum_{k=1}^{N(n)} a_{n,k} e^{2ik\pi \cdot} * \phi(t) \\ &= \lim_{n\to\infty} \sum_{k=1}^{N(n)} a_{n,k} e^{2ik\pi t} \int_{-\infty}^{\infty} e^{-2ik\pi s} \phi(s) ds \\ &= \lim_{n\to\infty} \sum_{k=1}^{N(n)} a_{n,k} e^{2ik\pi t} \hat{\phi}(2k\pi) \\ &= 0. \end{aligned}$$

This, by definition, shows that $sp(f) \subset \{m \in 2\pi\mathbf{Z} : f_m \neq 0\}$. Conversely, for $\lambda \in \{m \in 2\pi\mathbf{Z} : f_m \neq 0\}$ and for every sufficiently small positive ε we can choose a complex function $\varphi \in L^1(\mathbb{R})$ such that $\hat{\varphi}(\xi) = 1$, $\forall \xi \in [\lambda - \varepsilon/2, \lambda + \varepsilon/2]$ and $\hat{\varphi}(\xi) = 0$, $\forall \xi \notin [\lambda - \varepsilon, \lambda + \varepsilon]$. Repeating the above argument, we have

$$\begin{aligned} w(t) = f * \varphi(t) &= \lim_{n\to\infty} P_n(t) * \varphi(t) \\ &= \lim_{n\to\infty} \sum_{k=1}^{N(n)} a_{n,k} e^{2ik\pi t} \hat{\varphi}(2k\pi) \\ &= \lim_{n\to\infty} a_{n,k_0} e^{2ik_0\pi t}. \end{aligned} \tag{1.32}$$

Since $\lim_{n\to\infty} a_{n,k_0} = f_{k_0}$ this shows that $w \neq 0$. Thus, $\lambda \in sp(f)$. □

Exercise 8. Let $f(t) = ae^{-|t|}$. Show that $sp(f) = \mathbb{R}$.

Exercise 9. Let f be a continuous function with compact support. Show that

$$sp(f) = \{\rho \in \mathbb{R} : \int_{-\infty}^{\infty} f(t) e^{-i\rho t} dt \neq 0\}.$$

From this show that if f is positive and has compact support, then $0 \notin sp(f)$.

There is another way to approach the notion of spectrum of a bounded function via the Fourier- Carleman transform of a bounded function u defined by the formula

$$\hat{u}(\lambda) = \begin{cases} \int_0^\infty e^{-\lambda t} u(t)dt, (Re\lambda > 0); \\ \\ -\int_0^\infty e^{\lambda t} u(-t)dt, (Re\lambda < 0). \end{cases} \tag{1.33}$$

In fact, we will define the notion of Carleman spectrum of a bounded continuous function u as the set $\sigma(u)$ of all reals λ at which the Fourier-Carleman transform has a no holomorphic extension to any neighborhood of $i\lambda$. In fact, we compute $\sigma(u)$ in several simplest cases. Let $u(t) = ae^{i\lambda_0 t}, \lambda_0 \in \mathbb{R}, a \neq 0$. Then, for $Re\lambda > 0$

$$\begin{aligned} \hat{u}(\lambda) &= \int_0^\infty e^{-\lambda t} u(t)dt \\ &= \int_0^\infty e^{-\lambda t} a e^{i\lambda_0 t} dt \\ &= a\big(\lim_{t\to+\infty} \frac{1}{i\lambda_0 - \lambda} e^{(i\lambda_0 - \lambda)t} - \frac{1}{i\lambda_0 - \lambda}\big) \\ &= -\frac{a}{i\lambda_0 - \lambda}. \end{aligned}$$

Similarly, for $Re\lambda < 0$ we can compute $\hat{u}(\lambda)$ which is of the same form. Hence, $\hat{u}(\lambda)$ has holomorphic extension at any $i\xi \neq i\lambda_0$. Obviously, $\sigma(u) = \{\lambda_0\}$.

We consider now a more general case in which f is a τ-periodic continuous function.

Example 1.17. Let f be a $\mathbb{X}$-valued τ-periodic continuous function. Then

$$\sigma(f) = \{2\pi n/\tau \mid n \in \mathbb{Z}, \int_0^\tau e^{-i2\pi nt/\tau} f(t)dt \neq 0\}. \tag{1.34}$$

Proof. By definition, for $Re\lambda > 0$,

$$\begin{aligned}
\hat{f}(\lambda) &= \int_0^\infty e^{-\lambda t} f(t)dt \\
&= \sum_{n=1}^\infty \int_{(n-1)\tau}^{n\tau} e^{-\lambda t} f(t)dt \\
&= \sum_{n=1}^\infty \int_0^\tau e^{-\lambda(t+(n-1)\tau)} f(t+(n-1)\tau)dt \\
&= \sum_{n=1}^\infty e^{-(n-1)\lambda\tau} \int_0^\tau e^{-\lambda t} f(t)dt \\
&= \int_0^\tau e^{-\lambda t} f(t)dt \sum_{n=1}^\infty e^{-(n-1)\lambda\tau} \\
&= \int_0^\tau e^{-\lambda t} f(t)dt \frac{1}{1-e^{-\lambda\tau}}. \qquad (1.35)
\end{aligned}$$

Similarly, for $Re\lambda < 0$, (1.35) holds true as well. From (1.35) it is seen that if λ is such that $e^{-\lambda\tau} \neq 1$, i.e, $\lambda \neq 2\pi in/\tau$ for $n \in \mathbf{Z}$, then $\lambda \notin \sigma(f)$ because at this point $\hat{f}(\lambda)$ has a holomorphic extension. Moreover, at $\lambda_n = 2\pi in/\tau$, $\hat{f}(\lambda)$ has a holomorphic extension if and only if $\int_0^\tau e^{-2\pi nt/\tau} f(t)dt = 0$. This shows that (1.34) holds true. □

We have just shown that $\sigma(f) = sp(f)$ for periodic functions. In general, for bounded continuous functions they coincide with each other.

Theorem 1.12. *Under the notation as above, $sp(u)$ coincides with the set $\sigma(u)$.*

Proof. For the proof we refer the reader to Proposition 0.5, p.22 in [Pruss (91)]. □

Every definition of spectrum has its advantages. We will see this in the next chapter. Below we collect some main properties of the spectrum of a function, which we will need in the sequel.

Theorem 1.13. *Let $f, g_n \in BUC(\mathbb{R}, \mathbb{X}), n \in \mathbf{N}$ such that $g_n \to f$ as $n \to \infty$. Then*

(i) $sp(f)$ is closed,
(ii) $sp(f(\cdot + h)) = sp(f)$,
(iii) If $\alpha \in \mathbb{C}\backslash\{0\}$ $sp(\alpha f) = sp(f)$,

(iv) *If* $sp(g_n) \subset \Lambda$ *for all* $n \in \mathbf{N}$ *then* $sp(f) \subset \overline{\Lambda}$,
(v) *If* A *is a closed operator,* $f(t) \in D(A) \forall t \in \mathbb{R}$ *and* $Af(\cdot) \in BUC(\mathbb{R}, \mathbb{X})$, *then,* $sp(\mathcal{A}f) \subset sp(f)$,
(vi) $sp(\psi * f) \subset sp(f) \cap supp\mathcal{F}\psi, \forall \psi \in L^1(\mathbb{R})$.

Proof. The proofs of (i)-(iii) are straightforward. We refer the reader to Proposition 0.4, p. 20, Theorem 0.8 , p. 21 in [Vu (102)] and pp. 20-21 in [Pruss (91)] for the proofs of the remaining assertions. □

As a consequence of Theorem 1.13 we have the following:

Corollary 1.2. *Let* $\Lambda \subset \mathbb{R}$ *be closed. Then, the set*

$$\{f \in BUC(\mathbb{R}, \mathbb{X}) : sp(f) \subset \Lambda\} \tag{1.36}$$

is a closed subspace of $BUC(\mathbb{R}, \mathbb{X})$.

We consider the translation group $(S(t))_{t \in \mathbb{R}}$ on $BUC(\mathbb{R}, \mathbb{X})$. One of the frequently used properties of the spectrum of a function is the following:

Theorem 1.14. *Under the notation as above,*

$$i\ sp(u) = \sigma(\mathcal{D}_u), \tag{1.37}$$

where $\mathcal{D}_u$ *is the generator of the restriction of the group* $S(t)$ *to* $\mathcal{M}_u := \overline{span\{S(t)u, t \in \mathbb{R}\}}$.

Proof. For the proof see Theorem 8.19, p. 213 in [Davies (27)]. □

1.3.3 *Uniform Spectrum of a Bounded Function*

Notice that for every $\lambda \in \mathbb{C}$ with $\Re\lambda \neq 0$ and $f \in BC(\mathbb{R}, \mathbb{X})$ the function $\varphi_f(\lambda) : \mathbb{R} \ni t \mapsto \widehat{S(t)f}(\lambda) \in \mathbb{X}$ belongs to $\mathcal{M}_f \subset BC(\mathbb{R}, \mathbb{X})$. Moreover, $\varphi_f(\lambda)$ is analytic on $\mathbb{C} \backslash i\mathbb{R}$.

Definition 1.12. Let f be in $BC(\mathbb{R}, \mathbb{X})$. Then,

(1) $\alpha \in \mathbb{R}$ is said to be *uniformly regular* with respect to f if there exists a neighborhood $\mathcal{U}$ of $i\alpha$ in $\mathbb{C}$ such that the function $\varphi_f(\lambda)$, as a complex function of λ with $\Re\lambda \neq 0$, has an analytic continuation into $\mathcal{U}$.
(2) The set of $\xi \in \mathbb{R}$ such that ξ is not uniformly regular with respect to $f \in BC(\mathbb{R}, \mathbb{X})$ is called *uniform spectrum* of f and is denoted by $sp_u(f)$.

If $f \in BUC(\mathbb{R}, \mathbb{X})$, then $\alpha \in \mathbb{R}$ is uniformly regular if and only if it is regular with respect to f. In fact, this follows from the fact that for bounded uniformly continuous functions u, the identity (1.37) holds. Next, using the identity

$$R(\lambda, \mathcal{D}_u)u = \int_0^\infty e^{-(\lambda)\xi} S(\xi) u d\xi, \quad \Re\lambda \neq 0$$

we get the claim. For $f \in BC(\mathbb{R}, \mathbb{X})$, in general, the above (1.37) may not hold. We now study properties of uniform spectra of functions in $BC(\mathbb{R}, \mathbb{X})$.

Proposition 1.1. *Let $g, f, f_n \in BC(\mathbb{R}, \mathbb{X})$ such that $f_n \to f$ as $n \to \infty$ and let $\Lambda \subset \mathbb{R}$ be a closed subset. Then the following assertions hold:*

(i) $sp_u(f) = sp_u(f(h + \cdot))$;
(ii) $sp_u(\alpha f(\cdot)) \subset sp_u(f)$, $\alpha \in \mathbb{C}$;
(iii) $sp(f) \subset sp_u(f)$;
(iv) $sp_u(Bf(\cdot)) \subset sp_u(f)$, $B \in L(\mathbb{X})$;
(v) $sp_u(f + g) \subset sp_u(f) \cup sp_u(g)$;
(vi) $sp_u(f) \subset \Lambda$.

Proof. (i) - (v) are obvious from the definitions of spectrum and uniform spectrum. Now we prove (vi). Let $\rho_0 \notin \Lambda$. Since Λ is closed, there is a positive constant $r < dist(\rho_0, \Lambda)$. We can prove that since

$$\|\varphi_{f_n}(\lambda)\| \leq \frac{2\|f\|}{|\Re\lambda|}, \quad \forall \lambda \in \bar{B}_r(i\rho_0) \tag{1.38}$$

for sufficiently large $n \geq N$, one has

$$\|\varphi_{f_n}(\lambda)\| \leq \frac{4\|f\|}{3r}, \quad \forall \lambda \in \bar{B}_r(i\rho_0), n \geq N. \tag{1.39}$$

Obviously, for every fixed λ such that $\Re\lambda \neq 0$ we have $\varphi_{f_n}(\lambda) \to \varphi_f(\lambda)$. Now applying Vitali Theorem to the sequence of complex functions $\{\varphi_{f_n}\}$ we see that φ_{f_n} is convergent uniformly on $B_r(i\rho_0)$ to φ_f. This yields that φ_f is holomorphic on $B_r(i\rho_0)$, that is ρ_0 is a uniformly regular point with respect to f and $\rho_0 \notin sp_u(f)$. □

As an immediate consequence of (iii) of the above proposition, we have

Corollary 1.3. *For any closed subset $\Lambda \subset \mathbb{R}$, the set $\Lambda_u(\mathbb{X}) := \{f \in BC(\mathbb{R}, \mathbb{X}) : sp_u(f) \subset \Lambda\}$ is a closed subspace of $BC(\mathbb{R}, \mathbb{X})$ which is invariant under translations.*

The following result will be needed in the sequel.

Lemma 1.2. *Let Λ be a closed subset of $\mathbb{R}$ and let $\mathcal{D}_{\Lambda_u}$ be the differential operator acting on $\Lambda_u(\mathbb{X})$. Then we have*

$$\sigma(\mathcal{D}_{\Lambda_u}) = i\Lambda. \tag{1.40}$$

Proof. Since the function g_α defined by $g_\alpha(t) := e^{i\alpha t}x,\ \ \alpha \in \mathbb{R}, t \in \mathbb{R}, x \neq 0$, is in $\Lambda_u(\mathbb{X})$ and $sp_u(g_\alpha) = sp(g_\alpha) = \{\alpha\}$ we see that $i\alpha \in \sigma(\mathcal{D}_{\Lambda_u})$, that is, $i\Lambda \subset \sigma(\mathcal{D}_{\Lambda_u})$. Now we prove the converse. For $\beta \in \mathbb{R}\backslash\Lambda$ we consider the equation

$$i\beta g - g' = f, \quad f \in \Lambda_u(\mathbb{X}). \tag{1.41}$$

We will prove that (1.41) is uniquely solvable for every $f \in \Lambda_u(\mathbb{X})$. This equation has at most one solution. In fact, if g_1, g_2 are two solutions, then $g = g_1 - g_2$ is a solution of the homogeneous equation, that is for $f = 0$. Taking Carlemann transform of both sides of the correspoding equation we may see that $sp(g) \subset \{\beta\}$. Since $g \in \Lambda_u(\mathbb{X})$ we have $sp(g) \subset \Lambda$. Combining these facts we have $sp(g) = \emptyset$, that is $g = 0$.

Now we prove the existence of at least one solution to Eq. (1.41). For $\Re\lambda \neq 0$ Eq. (1.41) has a unique solution which is nothing but $\varphi_f(\lambda)$, so by definition,

$$\varphi_f(\lambda) = (\lambda - \mathcal{D}_f)^{-1} f, \quad \Re\lambda \neq 0.$$

Using a similar argument as in the proof of (iii) of Proposition 1.1 we can show that $(\lambda - \mathcal{D}_f)^{-1}u$ is bounded on $\bar{B}_r(i\beta)$ uniformly in $u \in span\{S(h)f, h \in \mathbb{R}\}, \|u\| \leq 1$ for certain positive constant r independent of u and λ. Since $i\beta$ is a limit point of $\sigma(\mathcal{D}_f)$, this boundedness yields in particular that $i\beta \in \rho(\mathcal{D}_f)$. Hence, there exists a unique solution $g \in \mathcal{M}_f \subset \Lambda_u(\mathbb{X})$ to (1.41). □

1.3.4 *Almost Periodic Functions*

1.3.4.1 *Definition and basic properties*

A subset $E \subset \mathbb{R}$ is said to be *relatively dense* if there exists a number $l > 0$ (*inclusion length*) such that every interval $[a, a + l]$ contains at least one point of E. Let f be a function on $\mathbb{R}$ taking values in a complex Banach space $\mathbb{X}$. f is said to be *almost periodic* if to every $\varepsilon > 0$ there corresponds a relatively dense set $T(\varepsilon, f)$ (*of ε-translations, or ε-periods*) such that

$$\sup_{t\in\mathbb{R}} \|f(t + \tau) - f(t)\| \leq \varepsilon, \ \forall \tau \in T(\varepsilon, f).$$

A typical example of an almost periodic function that is not periodic is the following:

Example 1.18.

$$f(t) = a\sin t + a\sin\sqrt{2}t\ ,\ \forall\, 0 \neq a \in \mathbb{X}. \tag{1.42}$$

In general, the function

$$f(t) = ae^{it} + be^{i\sqrt{2}t},\ \ a, b \in \mathbb{X}, a \neq 0, b \neq 0,$$

is an almost periodic one that is not periodic.

Proof. We will make use of only the definitions and the following fact from the elementary mathematics: for every constant $\varepsilon > 0$ there exists a positive integer $N(\varepsilon)$ such that $N\sqrt{2} - [N\sqrt{2}] < \varepsilon$, where $[r]$ denotes the integer part of the real number r. By this fact, for a positive ε there is an interval $(2M\pi - \alpha, 2M\pi + \alpha)$, where M is an integer and $\alpha > 0$, such that

$$\|be^{i\sqrt{2}(t+\tau)} - be^{i\sqrt{2}t}\| < \frac{\varepsilon}{2},\ \forall t, \tau \in (2M\pi - \alpha, 2M\pi + \alpha).$$

Hence, for sufficiently small α and $l = 2M\pi$ every inteval of length l contains at least an ε-period of the function f. This shows that f is almost periodic. Now we are going to prove that f is not periodic. In fact, suppose to the contrary that f is periodic with period T. By this assumption, the function

$$g(t) := ae^{i(t+T)} + be^{i\sqrt{2}(t+T)} - ae^{it} - be^{i\sqrt{2}t} = 0\ \forall t \in \mathbb{R},$$

Thus,

$$\begin{aligned} 0 = \int_0^{2\pi} g(t)dt &= b\int_0^{2\pi} (e^{i\sqrt{2}(t+T)} - e^{i\sqrt{2}t})dt \\ &= \frac{1}{i\sqrt{2}}(e^{i\sqrt{2}2\pi} - 1)(e^{i\sqrt{2}T} - 1). \end{aligned} \tag{1.43}$$

This shows that $T/\sqrt{2}$ must be rational. Similarly, we can show that T is rational. This leads to a contradiction showing that f is not periodic. □

Generally, the sum of almost periodic functions are an almost periodic function.

Example 1.19. All trigonometric polynomials

$$P(t) = \sum_{k=1}^{n} a_k e^{i\lambda_k t},\ (a_k \in \mathbb{X}, \lambda_k \in \mathbb{R})$$

are almost periodic.

We collect some basic properties of an almost periodic function in the following:

Theorem 1.15. *Let f and $f_n, n \in \mathbb{R}$ be almost periodic functions with values in $\mathbb{X}$. Then the following assertions hold true:*

(1) The range of f is precompact, i.e., the set $\overline{\{f(t), t \in \mathbb{R}\}}$ is a compact subset of $\mathbb{X}$, so f is bounded;
(2) f is uniformly continuous on $\mathbb{R}$;
(3) If $f_n \to g$ as $n \to \infty$ uniformly, then g is almost periodic;
(4) If f' is uniformly continuous, then f' is almost periodic.

Proof. For the proof see e.g. pp. 5-6 [Amerio and Prouse (3)]. □

As a consequence of Theorem 1.15 the space of all almost periodic functions taking values in $\mathbb{X}$ with sup-norm is a Banach space which will be denoted by $AP(\mathbb{X})$. For almost periodic functions the following criterion holds (*Bochner's criterion*):

Theorem 1.16. *Let f be a continuous function taking values in $\mathbb{X}$. Then f is almost periodic if and only if given a sequence $\{c_n\}_{n\in\mathbf{N}}$ there exists a subsequence $\{c_{n_k}\}_{k\in\mathbf{N}}$ such that the sequence $\{f(t + c_{n_k})\}_{k\in\mathbf{N}}$ converges uniformly.*

Proof. For the proof see e.g. p. 9 in [Amerio and Prouse (3)]. □

Exercise 10. Let $f \in BUC(\mathbb{R}, \mathbb{X})$ such that $e^{isp(f)}$ is finite. Show that f is of the form (1.44), so f is almost periodic.

Proof. By Theorem 1.14

$$e^{isp(f)} = e^{\sigma(\mathcal{D}_f)}.$$

On the other hand, by the Weak Spectral Mapping Theorem,

$$\overline{e^{\sigma(\mathcal{D}_f)}} = \sigma(S(1)|_{\mathcal{M}_f}).$$

Hence, using Riesz Integral we can decompose $\mathcal{M}_f$ into the direct sum of finite closed subspaces $\mathcal{M}_1, ..., \mathcal{M}_k$ invariant under the translation group $(S(t))_{t\in\mathbb{R}}$. Moreover, the spectrum of $S(1)$ restricted to every subspace consists of only one point. By Gelfand Theorem, $S(1)$ should have the following form: $S(1)|_{\mathcal{M}_j} = e^{i\lambda_j}I,\ \lambda_j \in \mathbb{R}$. It is easy to see that if, by the above decomposition, $f = f_1 + ... + f_k$ then the function $g_j(t) := e^{-i\lambda_j t} f_j(t)$ satisfies

$$\begin{aligned} g_j(t+1) &= e^{-i\lambda_j(t+1)} f_j(t+1) \\ &= e^{-i\lambda_j t} e^{-i\lambda_j} e^{i\lambda_j} f_j(t) \\ &= e^{-i\lambda_j t} e^{i\lambda_j} f_j(t) \\ &= g_j(t), \ \forall t \in \mathbb{R}, \ j = 1, 2, ..., k. \end{aligned}$$

Finally,

$$f(t) = \sum_{j=1}^{k} e^{i\lambda t} g_j(t), \tag{1.44}$$

where $g_j(\cdot)$ is 1-periodic. This shows that f is almost periodic. □

1.3.5 *Sprectrum of an Almost Periodic Function*

There is a natural extension of the notion of Fourier exponents of periodic functions to almost periodic functions. In fact, if f is almost periodic function taking values in $\mathbb{X}$, then for every $\lambda \in \mathbb{R}$ the average

$$a(f, \lambda) := \lim_{T \to \infty} \frac{1}{2T} \int_{-T}^{T} e^{-i\lambda t} f(t) dt$$

exists and is different from 0 at most at countably many points λ. The set $\{\lambda \in \mathbb{R} : a(f, \lambda) \neq 0\}$ is called *Bohr spectrum* of f which will be denoted by $\sigma_b(f)$. The following Approximation Theorem of almost periodic functions holds

Theorem 1.17. *(Approximation Theorem) Let f be an almost periodic function. Then for every $\varepsilon > 0$ there exists a trigonometric polynomial*

$$P_\varepsilon(t) = \sum_{j=1}^{N} a_j e^{i\lambda_j t}, \ a_j \in \mathbb{X}, \lambda_j \in \sigma_b(f)$$

such that

$$\sup_{t \in \mathbb{R}} \|f(t) - P_\varepsilon(t)\| < \varepsilon.$$

Proof. For the proof see e.g. pp. 17-24 in [Levitan and Zhikov (58)]. □

Remark 1.3. The trigonometric polynomials $P_\varepsilon(t)$ in Theorem 1.17 can be chosen as an element of the space

$$\mathcal{M}_f := \overline{span\{S(\tau)f, \tau \in \mathbb{R}\}}$$

(see p. 29 in [Levitan and Zhikov (58)]. Moreover, without loss of generality by assuming that $\sigma_b(f) = \{\lambda_1, \lambda_2, \cdots\}$ one can choose a sequence of trigonometric polynomials, called *trigonometric polynomials of Bochner-Fejer*, approximating f such that

$$P_m(t) = \sum_{j=1}^{N(m)} \gamma_{m,j} a(\lambda_j, f) e^{i\lambda_j t}, m \in \mathbf{N},$$

where $\lim_{m\to\infty} \gamma_{m,j} = 1$. As a consequence we have:

Corollary 1.4. *Let f be almost periodic. Then*

$$\mathcal{M}_f = \overline{span\{a(\lambda, f)e^{i\lambda\cdot}, \lambda \in \sigma_b(f)\}}.$$

Proof. By Theorem 1.17,

$$\mathcal{M}_f \subset \overline{span\{a(f, \lambda)e^{i\lambda\cdot}, \lambda \in \sigma_b(f)\}}.$$

On the other hand, it is easy to prove by induction that if P is any trigonometric polynomial with different exponents $\{\lambda_1, \cdots, \lambda_k\}$, such that

$$P(t) = \sum_{j=1}^{k} x_j e^{i\lambda_k t},$$

then $x_j e^{i\lambda_j} \in \mathcal{M}_P$, $\forall j = 1, \cdots, k$. Hence by Remark 1.3, obviously, $a(\lambda_j, f)e^{i\lambda_j} \in \mathcal{M}_f$, $\forall j \in \mathbf{N}$. □

The relation between the spectrum of an almost periodic function f and its Bohr spectrum is stated in the following:

Proposition 1.2. *If f is an almost periodic function, then $sp(f) = \overline{\sigma_b(f)}$.*

Proof. Let $\lambda \in \sigma_b(f)$. Then there is a $x \in \mathbb{X}$ such that $xe^{i\lambda\cdot} \in \mathcal{M}_f$. Obviously, $\lambda \in \sigma(\mathcal{D}|_{\mathcal{M}_f})$. By Theorem 1.14 $\lambda \in sp(f)$. Conversely, by Theorem 1.17, f can be approximated by a sequence of trigonometric polynomials with exponents contained in $\sigma_b(f)$. In view of Theorem 1.13 $sp(f) \subset \overline{\sigma_b(f)}$. □

1.3.6 *A Spectral Criterion for Almost Periodicity of a Function*

Suppose that we know beforehand that $f \in BUC(\mathbb{R}, \mathbb{X})$. It is often possible to establish the almost periodicity of this function starting from certain *a priori* information about its spectrum.

Theorem 1.18. *Let $\mathcal{E}$ and $\mathcal{G}$ be closed, translation invariant subspaces of $BUC(\mathbb{R}, \mathbb{X})$ and suppose that*

(1) $\mathcal{G} \subset \mathcal{E}$;
(2) $\mathcal{G}$ *contains all constant functions which belong to* $\mathcal{E}$;
(3) $\mathcal{E}$ *and* $\mathcal{G}$ *are invariant under multiplications by* $e^{i\xi \cdot}$ *for all* $\xi \in \mathbb{R}$;
(4) *whenever* $f \in \mathcal{G}$ *and* $F \in \mathcal{E}$, *where* $F(t) = \int_0^t f(s)ds$, *then* $F \in \mathcal{G}$.

Let $u \in \mathcal{E}$ *have countable reduced spectrum*

$$sp_{\mathcal{G}} := \{\xi \in \mathbb{R} : \forall \varepsilon > 0 \exists f \in L^1(\mathbb{R}) \text{ such that } \\ supp\mathcal{F}f \subset (\xi - \varepsilon, \xi + \varepsilon) \text{ and } f * u \notin \mathcal{G}\}.$$

Then $u \in \mathcal{G}$.

Proof. For the proof see p. 371 in [Arendt and Batty (5)]. □

Remark 1.4. In the case where $\mathcal{G} = AP(\mathbb{X})$ the condition iv) in Theorem 1.18 can be replaced by the condition that $\mathbb{X}$ does not contain c_0 (see Proposition 3.1, p. 369 in [Arendt and Batty (5)]). Another alternative of the condition iv) is the total ergodicity of u which is defined as follows: $u \in BUC(\mathbb{R}, \mathbb{X})$ is called *totally ergodic* if

$$M_\eta u := \lim_{\tau \to \infty} \frac{1}{2\tau} \int_{-\tau}^{\tau} e^{i\eta s} S(s) ds$$

exists in $BUC(\mathbb{R}, \mathbb{X})$ for all $\eta \in \mathbb{R}$. From this remark the following example is obvious:

Example 1.20. A function $f \in BUC(\mathbb{R}, \mathbb{X})$ is 2π-periodic if and only if $sp(f) \subset 2\pi\mathbf{Z}$.

1.3.7 *Almost Automorphic Functions*

Definition and Basic Properties. A function $f \in C(\mathbb{R}, \mathbb{X})$ is said to be *almost automorphic* if for any sequence of real numbers (s_n'), there exists a subsequence (s_n) such that

$$\lim_{m \to \infty} \lim_{n \to \infty} f(t + s_n - s_m) = f(t) \tag{1.45}$$

for any $t \in \mathbb{R}$.

The limit in (1.45) means

$$g(t) = \lim_{n \to \infty} f(t + s_s) \tag{1.46}$$

is well-defined for each $t \in \mathbb{R}$ and

$$f(t) = \lim_{n\to\infty} g(t - s_n) \tag{1.47}$$

for each $t \in \mathbb{R}$.

Remark 1.5. Because of pointwise convergence the function g is measurable but not necessarily continous.

Remark 1.6. It is also clear from the definition above that constant functions and continuous almost periodic functions are almost automorphic.

If the limit in (1.41) is uniform on any compact subset $K \subset \mathbb{R}$, we say that f is compact almost automorphic.

Theorem 1.19.
Assume that f, f_1, and f_2 are almost automorphic functions taking values in a Banach space $\mathbb{X}$, ϕ is a scalar almost automorphic function, and λ is any scalar, then the following hold true.

(i) λf *and* $f_1 + f_2$ *are almost automorphic,*
(ii) $f_\tau(t) := f(t+\tau)$, $t \in \mathbb{R}$ *is almost automorphic;*
(iii) $\bar{f}(t) := f(-t)$, $t \in \mathbb{R}$ *is almost automorphic;*
(iv) The Range R_f *of* f *is precompact, so* f *is bounded;*
(v) The function $t \mapsto \phi(t)f(t)$ *is almost automorphic.*

Proof. See Theorems 2.1.3 and 2.1.4 in [N'Guérékata (79)], for the proofs of (i)-(iv). The proof of (v) is straightforward, and is left to the reader. □

Theorem 1.20. *If $\{f_n\}$ is a sequence of almost automorphic $\mathbb{X}$-valued functions such that $f_n \to f$ uniformly on $\mathbb{R}$, then f is almost automorphic.*

Proof. See Theorem 2.1.10 in [N'Guérékata (79)], for proof. □

Remark 1.7. If we equip $AA(\mathbb{X})$, the space of almost automorphic functions with the sup norm

$$\|f\|_\infty = \sup_{t\in\mathbb{R}} \|f(t)\|$$

then it turns out to be a Banach space. If we denote $KAA(\mathbb{X})$, the space of compact almost automorphic $\mathbb{X}$-valued functions, then we have

$$AP(\mathbb{X}) \subset KAA(\mathbb{X}) \subset AA(\mathbb{X}) \subset BC(\mathbb{R}, \mathbb{X})\,.$$

Theorem 1.21. *If $f \in AA(\mathbb{X})$, then*

i) $\|f\|_\infty = \|g\|_\infty$

ii) $R_g \subset \overline{R_f}$

where g is the function defined in (1.41)-(1.42).

Proof. i) Using (1.41) we may write

$$\|g(t)\| \leq \|g(t) - f(t+s_n)\| + \|f(t+s_n)\|$$

and choosing n large enough, we get

$$\|g(t)\| < \varepsilon + \sup_{\sigma \in \mathbb{R}} \|f(\sigma)\|$$

Hence

$$\sup_{t \in \mathbb{R}} \|g(t)\| \leq \sup_{t \in \mathbb{R}} \|f(t)\|$$

Similarly by (1.42), we obtain

$$\sup_{t \in \mathbb{R}} \|f(t)\| \leq \sup_{t \in \mathbb{R}} \|g(t)\| \tag{1.48}$$

which proves the theorem.

ii) This statement is straight forward. □

Theorem 1.22. *If $f \in AA(\mathbb{X})$ and its derivative f' exists and is uniformly continuous on $\mathbb{R}$, then $f' \in AA(\mathbb{X})$.*

Proof. It suffices observe that for each $n \in \mathbb{N}$, $n(f(t+\frac{1}{n}) - f(t))$ is an almost automorphic function and the sequence of these functions converges uniformly to f' on $\mathbb{R}$ (see Theorem 2.4.1 in [N'Guérékata (79)] for a detailed proof). □

Theorem 1.23. *Let us define $F : \mathbb{R} \mapsto \mathbb{X}$ by $F(t) = \int_0^t f(s)ds$ where $f \in AA(\mathbb{X})$. Then $F \in AA(\mathbb{X})$ iff $R_F = \{F(t)/t \in \mathbb{R}\}$ is precompact.*

Before we prove the Theorem, let us introduce some useful notations (due to S. Bochner).

Remark 1.8. If $f : \mathbb{R} \to X$ is a function and a sequence of real numbers $s = (s_n)$ is such that we have

$$\lim_{n \to \infty} f(t+s_n) = g(t), \quad \text{pointwise on } \mathbb{R},$$

we will write $T_s f = g$.

Remark 1.9.

i) T_s is a linear operator.
Indeed, given a fixed sequence $s = (s_n) \subset \mathbb{R}$, the domain of T_s is $D(T_s) = \{f : \mathbb{R} \to X \,/\, T_s f \text{ exists}\}$. $D(T_s)$ is a linear set for if $f, f_1, f_2 \in D(T_s)$, then $f_1 + f_2 \in D(T_s)$ and $\lambda f \in D(T_s)$ for any scalar λ. And obviously, $T_s(f_1 + f_2) = T_s f_1 + T_s f_2$ and $T_s(\lambda f) = \lambda T_s f$.

ii) Let us write $-s = (-s_n)$ and suppose that $f \in D(T_s)$ and $T_s f \in D(T_{-s})$. Then the product operator $A_s = T_{-s} T_s f$ is well defined. It is easy to verify that A_s is also a linear operator.

iii) A_s maps bounded functions into bounded functions, and for almost automorphic functions f, we get $A_s f = f$.

We are now ready to prove the previous Theorem:

Proof. It suffices to prove that $F(t)$ is almost automorphic if R_F is precompact. Let (s''_n) be a sequence of real numbers. Then there exists a subsequence (s'_n) such that

$$\lim_{n\to\infty} f(t + s'_n) = g(t)$$

and

$$\lim_{n\to\infty} g(t - s'_n) = f(t),$$

pointwise on $\mathbb{R}$, and

$$\lim_{n\to\infty} F(s'_n) = \alpha_1,$$

for some vector $\alpha_1 \in X$.

We get for every $t \in \mathbb{R}$:

$$F(t + s'_n) = \int_0^{t+s'_n} f(r)\,dr = \int_0^{s'_n} f(r)\,dr + \int_{s'_n}^{t+s'_n} f(r)\,dr$$

$$= F(s'_n) + \int_{s'_n}^{t+s'_n} f(s)\,dr.$$

Using the substitution $\sigma = r - s'_n$, we obtain

$$F(t + s'_n) = F(s'_n) + \int_0^t f(\sigma + s'_n)\,d\sigma.$$

If we apply the Lebesgue's dominated convergence theorem, we obtain

$$\lim_{n\to\infty} F(t + s'_n) = \alpha_1 + \int_0^t g(\dot{\sigma})\,d\sigma$$

for each $t \in \mathbb{R}$.

Let us observe that the range of the function $G(t) = \alpha_1 + \int_0^t g(r)\, dr$ is also precompact and

$$\sup_{t \in \mathbb{R}} \|G(t)\| = \sup_{t \in \mathbb{R}} \|F(t)\|$$

by Theorem 1.21 i), so that we can extract a subsequence (s_n) of (s'_n) such that

$$\lim_{n \to \infty} G(-s_n) = \alpha_2,$$

for some $\alpha_2 \in X$.

Now we can write

$$G(t - s_n) = G(-s_n) + \int_0^t g(r - s_n)\, dr$$

so that

$$\lim_{n \to \infty} G(t - s_n) = \alpha_2 + \int_0^t f(r)\, dr = \alpha_2 + F(t).$$

Let us prove now that $\alpha_2 = \theta$.

Using the notation above we get

$$A_s F = \alpha_2 + F, \quad \text{where} \quad s = (s_n).$$

Now it is easy to observe that F as well as α_2 belong to the domain of A_s; therefore $A_s F$ also is in the domain of A_s and we deduce the equation

$$A_s^2 F = A_s \alpha_2 + A_s F = \alpha_2 + \alpha_2 + F = 2\alpha_2 + F$$

We can continue indefinitely the process to get

$$A_s^n F = n\alpha_2 + F, \quad \forall n = 1, 2, \ldots.$$

But we have

$$\sup_{t \in \mathbb{R}} \|A_s^n F(t)\| \leq \sup_{t \in \mathbb{R}} \|F(t)\|$$

and $F(t)$ is a bounded function.

This leads to a contradiction if $\alpha_2 \neq 0$. Hence, $\alpha_2 = 0$ and $A_s F = F$; so $F \in AA(\mathbb{X})$.

The proof is complete. □

Remark 1.10. If $\mathbb{X}$ is a uniformly convex Banach space, the assumption on R_F can be weakened. Indeed, the result holds true if R_F is bounded (see, Theorem 2.4.4 and Theorem 2.4.6 in [N'Guérékata (79)]).

We can recall this other important result:

Theorem 1.24. *Let* $(T(t))_{t\in\mathbb{R}}$ *be a* C_0*-group of bounded linear operators, and assume that the function* $x(t) = T(t)x_0 : \mathbb{R} \mapsto \mathbb{X}$*, where* $x(0) = x_0 \in \mathbb{X}$ *is almost automorphic. Then either* $\inf_{t\in\mathbb{R}} \|x(t)\| > 0$*, or* $x(t) = 0$ *for every* $t \in \mathbb{R}$.

Proof. Assume that $\inf_{t\in\mathbb{R}} \|x(t)\| = 0$. Then there exists a minimizing sequence (s_n') such that $\|x(s_n')\| \mapsto 0$ as $n \mapsto \infty$. Since $x(t)$ is almost automorphic, there exists a subsequence $(s_n) \subset (s_n')$ such that

$$\lim_{n\to\infty} x(t + s_n) = y(t)$$

exists for every $t \in \mathbb{R}$, and

$$\lim_{n\to\infty} y(t - s_n) = x(t)$$

for every $t \in \mathbb{R}$.
We have

$$x(t + s_n) = T(t + s_n) - T(t)T(s_n)x_0 = T(t)x(s_n) .$$

So,

$$\|y(t)\| = \lim_{n\to\infty} \|T(t)x(s_n)\| \leq \lim_{n\to\infty} \|T(t)\| \|x(s_n)\| = 0$$

for every $t \in \mathbb{R}$. We infer that $x(t) = 0$ for every $t \in \mathbb{R}$. The theorem is proved. □

Definition 1.13. A function $f \in C(\mathbb{R}^+, \mathbb{X})$ is said to be asymptotically almost automorphic, if there exists $g \in AA(\mathbb{X})$ and $h \in C(\mathbb{R}^+, \mathbb{X})$ with the property that $\lim_{t\to\infty} \|h(t)\| = 0$, such that

$$f(t) = g(t) + h(t), \quad t \in \mathbb{R}^+ . \tag{1.49}$$

The functions g and h are called respectively the principal and the corrective terms of f.

Theorem 1.25. *If* f *is asymptotically almost automorphic then its principal and corrective terms are uniquely determined.*

Proof. See Theorem 2.5.4 in [N'Guérékata (79)]. □

Exercise 11. Prove that every asymptotically almost function is bounded over $\mathbb{R}^+$.

Exercise 12. Let $f \in C(\mathbb{R}^+, \mathbb{X})$ and $\nu \in C(\mathbb{R}^+, \mathbb{C})$ be asymptotically almost automorphic. Show that $f_\tau(t) := f(t+\tau)$, for a fixed $\tau \in \mathbb{R}^+$ and $(\nu f)(t) = \nu(t)f(t)$ are also asymptotically almost automorphic.

Exercise 13. Let $AAA(\mathbb{X})$ be the space of asymptotically almost automorphic functions with the norm $||f||_{AAA(\mathbb{X})} = ||g||_{AA(\mathbb{X})} + ||h||_{C(\mathbb{R}^+,\mathbb{X})}$, where g and h are respectiveley the prinicipal and the corrective terms of f. Show that if (f_n) is a sequence of functions in $AAA(\mathbb{X})$ that converges uniformly to f, then $f \in AAA(\mathbb{X})$.

Chapter 2

Stability and Exponential Dichotomy of Solutions of Homogeneous Equations

This chapter is devoted to the study of the behavior of homogeneous equations of the form $u'(t) = A(t)u(t), t \in \mathbb{R}$. The exponential dichotomy and stability will be characterized in terms of Perron's criteria.

2.1 Perron Theorem

In this section we will extend the Perron Theorem for ordinary differential equations to equations in Banach spaces. Throughout this section we consider only processes defined on the whole line. We denote $\Delta := \{(t,s) \in \mathbb{R} : t \geq s\}$.

Definition 2.1. A family of bounded linear operators $(U(t,s))_{t\geq s}, t, s \in \mathbb{R}$ from a Banach space $\mathbb{X}$ to itself is called *a strongly continuous evolutionary process* if the following conditions are satisfied

(1) $U(t,t) = I$ for all $t \in \mathbb{R}$,
(2) $U(t,s)U(s,r) = U(t,r)$ for all $t \geq s \geq r$,
(3) The map $\Delta \ni (t,s) \mapsto U(t,s)x$ is continuous for every fixed $x \in \mathbb{X}$,
(4) $\|U(t,s)\| \leq Ne^{\omega(t-s)}$ for some positive N, ω independent of $(t,s) \in \Delta$.

Definition 2.2. Let $(U(t,s))_{t\geq s}$ be a strongly continuous evolutionary process on a Banach space $\mathbb{X}$. $(U(t,s))_{t\geq s}$ is said to have an *exponential dichotomy* if there exist a family of projections $Q(t), t \in \mathbb{R}$ and positive constants M, α such that the following conditions are satisfied:

(1) For every fixed $x \in \mathbf{X}$ the map $t \mapsto Q(t)x$ is continuous,
(2) $Q(t)U(t,s) = U(t,s)Q(s), \forall (t,s) \in \Delta$,
(3) $\|U(t,s)x\| \leq Me^{-\alpha(t-s)}\|x\|, \forall (t,s) \in \Delta, x \in KerQ(s)$,

(4) $\|U(t,s)y\| \geq M^{-1}e^{\alpha(t-s)}\|y\|, \forall (s,t) \in \Delta, y \in ImQ(s)$,
(5) $U(t,s)|_{ImQ(s)}$ is an isomorphism from $ImQ(s)$ onto $ImQ(t)$, $\forall (t,s) \in \Delta$.

In this section we will establish a relation between the exponential dichotomy of a given strongly continuous process $(U(t,s))_{t\geq s}$ and the unique solvability of the following integral equation in appropriate function spaces:

$$u(t) = U(t,s)u(s) + \int_s^t U(t,\xi)f(\xi)d\xi, \quad \forall (t,s) \in \Delta. \tag{2.1}$$

Next, we define an operator $\mathcal{L} : D(\mathcal{L}) \subset C_0(\mathbb{R},\mathbb{X}) \to C_0(\mathbb{R},\mathbb{X})$ as follows: $u \in C_0(\mathbb{R},\mathbb{X})$ if there exists a function $f \in C_0(\mathbb{R},\mathbb{X})$ such that (2.1) holds. For $u \in D(\mathcal{L})$ we define $\mathcal{L}u = f$.

Lemma 2.1. *The operator $\mathcal{L}$ is a well-defined, closed and linear operator.*

Proof. To prove the well-definedness of the operator we suppose that

$$u(t) := U(t,s)u(s) + \int_s^t U(t,\xi)g(\xi)d\xi, \quad \forall (t,s) \in \Delta$$

$$u(t) := U(t,s)u(s) + \int_s^t U(t,\xi)f(\xi)d\xi, \quad \forall (t,s) \in \Delta.$$

Then,

$$\int_s^t U(t,\xi)g(\xi)d\xi = \int_s^t U(t,\xi)f(\xi)d\xi.$$

Hence,

$$\int_s^t U(t,\xi)[f(\xi) - g(\xi)]d\xi = 0.$$

Thus,

$$\frac{1}{(t-s)}\int_s^t U(t,\xi)[f(\xi) - g(\xi)]d\xi = 0.$$

Letting $t - s \to 0$, by the strong continuity of the process $(U(t,s))_{t\geq s}$ we obtain that

$$f(t) = g(t), \quad t \in \mathbb{R}.$$

So, $\mathcal{L}$ is well-defined.

Let $\{v_n\}$ be a sequence in $D(\mathcal{L})$, such that $\lim_{n\to\infty}\|v_n - v\| = 0$ for some $v \in C_0(\mathbb{R},\mathbb{X})$ and $\exists f \in C_0(\mathbb{R},\mathbb{X})$ such that $\lim_{n\to\infty}\|\mathcal{L}v_n - f\| = 0$. Now we prove that $v \in D(\mathcal{L})$ and $\mathcal{L}v = f$. In fact, we have

$$v_n(t) = U(t,s)v_n(s) + \int_s^t U(t,\xi)\mathcal{L}v_n(\xi)d\xi, \quad (t,s) \in \Delta. \tag{2.2}$$

For fixed $(t,s) \in \Delta$, we have

$$\|\int_s^t U(t,\xi)\mathcal{L}v_n(\xi)d\xi - \int_s^t U(t,\xi)f(\xi)d\xi\| \le \int_s^t \|U(t,\xi)\|.\|\mathcal{L}v_n(\xi) - f(\xi)\|d\xi$$

$$\le N_1 \int_s^t \|\mathcal{L}v_n(\xi) - f(\xi)\|d\xi \le N_1(t-s)\|\mathcal{L}v_n - f\|$$

From this we obtain

$$\lim_{n\to\infty} \|\int_s^t U(t,\xi)\mathcal{L}v_n(\xi)d\xi - \int_s^t U(t,\xi)f(\xi)d\xi\| = 0. \tag{2.3}$$

This yields

$$v(t) = U(t,s)v(s) + \int_s^t U(t,\xi)f(\xi)d\xi, \quad (t,s) \in \Delta.$$

Therefore, $v \in D(\mathcal{L})$ and $\mathcal{L}v = f$. The linearity of the operator $\mathcal{L}$ can be easily shown. □

Theorem 2.1. *Let $(U(t,s))_{t\ge s}$ be a given strongly continuous evolutionary process. Then the following assertions are equivalent:*

(i) The process $(U(t,s))_{t\ge s}$ has an exponential dichotomy;
(ii) For every given $f \in C_0(\mathbb{R},\mathbb{X})$ the integral equation (2.1) has a unique solution $x_f \in C_0(\mathbb{R},\mathbb{X})$;
(iii) For every given bounded and continuous function f the integral equation (2.1) has a unique bounded solution.

Proof. We first prove the equivalence between (i) and (ii). The equivalence between (i) and (ii) can be established in the same way.
"(i) ⇒ (ii)": Let $P(t) := I - Q(t), \quad t \in \mathbb{R}$, where $Q(t)$ are determined from the exponential dichotomy of the process $(U(t,s))_{t\ge s}$. We now show that the equation $\mathcal{L}u = 0$ has the only trivial solution $u = 0$. In fact, by the definition of exponential dichotomy (Definition 2.2), we have

$$\begin{aligned}\|P(t)u(t)\| &= \|U(t,s)P(s)u(s)\| \\ &\le Me^{-\alpha(t-s)}\|u(s)\| \\ &\le Me^{-\alpha(t-s)}\|u\|, \quad \forall t \ge s.\end{aligned}$$

Hence, letting $s \to -\infty$ we get $P(t)u(t) = 0, \ \forall t \in \mathbb{R}$. Similarly, we can show $Q(t)u(t) = 0, \ \forall t \in \mathbb{R}$. This yields that $u(t) = 0, \ \forall t \in \mathbb{R}$.

Next, we define a so-called Green function

$$G(t,s) = \begin{cases} U(t,s)P(s), & t \geq s \\ -U(t,s)Q(s), & t < s. \end{cases}$$

By the exponential dichotomy, we see that there are positive constants K, α such that

$$\|G(t,s)\| \leq Ke^{-\alpha|t-s|}, \quad \forall t,s \in \mathbb{R}.$$

We claim that the function

$$u(t) = \int_{-\infty}^{\infty} G(t,\xi)f(\xi)d\xi, \quad t \in \mathbb{R}$$

is a (unique) bounded continuous solution to the integral equation (2.1). In fact, for any $t \geq s$ we have

$$\begin{aligned} u(t) &= \int_{-\infty}^{t} U(t,\xi)P(\xi)f(\xi)d\xi - \int_{t}^{+\infty} U(t,\xi)Q(\xi)f(\xi)d\xi \\ &= U(t,s)\left(\int_{-\infty}^{s} U(s,\xi)P(\xi)f(\xi)d\xi - \int_{s}^{+\infty} U(s,\xi)Q(\xi)f(\xi)d\xi\right) \\ &\quad + \int_{s}^{t} U(t,\xi)P(\xi)f(\xi)d\xi + \int_{s}^{t} U(t,\xi)(Q(\xi)f(\xi)d\xi \\ &= U(t,s)u(s) + \int_{s}^{t} U(t,\xi)f(\xi)d\xi. \end{aligned}$$

On the other hand, we have

$$\begin{aligned} \|u(t)\| &= \|\int_{-\infty}^{+\infty} G(t,s)f(s)ds\| \\ &\leq \int_{-\infty}^{+\infty} Ke^{-\alpha|t-s|}\|f(s)\|ds \\ &\leq \int_{-\infty}^{+\infty} Ke^{-\alpha|\eta|}\|f(t-\eta)\|d\eta. \end{aligned}$$

An easy computation shows that as $f \in C_0(\mathbb{R}, \mathbb{X})$, the above estimate yields $u \in C_0(\mathbb{R}, \mathbb{X})$.

Now we prove the implication "$(ii) \Rightarrow (i)$". First, we notice that by the above lemma and the Closed Graph Theorem (ii) yields that there exists a positive constant k such that

$$\|\mathcal{L}^{-1}f\| \leq c\|f\|, \quad \forall f \in C_0(\mathbb{R}, \mathbb{X}). \tag{2.4}$$

To proceed we need several lemmas:

Lemma 2.2. *Let (ii) hold. Then there exist positive constants $M, \beta > 0$ depending only on c such that*

$$\|U(t,t_0)x\| \le Me^{-\beta(t-s)}\|U(s,t_0)x\|, \quad \forall t \ge s \ge t_0, \tag{2.5}$$

provided that

$$\lim_{t\to+\infty} U(t,t_0)x = 0. \tag{2.6}$$

Proof. Without loss of generality we may assume that $x \neq 0$. Set $u(t) := U(t,t_0)x$, $\forall t \ge t_0$ and $t_1 := \sup\{t \ge t_0 : u(t) \neq 0\}$. Obviously, by definition, $u(t) \neq 0$ for all $t \in [t_0, t_1)$ and $u(t) = 0$ for all $t \ge t_1$. Next, we define a smooth function $\phi_n(t)$ for every $n \in \mathbb{N}$ with compact $supp\phi_n$ in $[t_0, t_1)$ such that $0 \le \phi_n(t) \le 1$, $\phi_n(t) = 1$ for all $t \in [t_0 + 1/n, \min(n, t_1 - 1/n)]$. Set

$$f(t) = \frac{\phi_n(t)}{\|u(t)\|}u(t), \quad t \in \mathbb{R}.$$

It is clear that f is a well-defined continuous function on $\mathbb{R}$ and $f \in C_0(\mathbb{R}, \mathbb{X})$. Consider the function

$$x_f(t) := u(t)\int_{t_0}^{t} \frac{\phi_n(s)}{\|u(s)\|}ds, \quad t \ge t_0.$$

By the definition of ϕ_n we see that x_f can be naturally extended to the whole line as a continuous function by setting $x_f(t) = 0$ for all $t \notin [t_0, \min(n, t_1)]$. Consequently, we have

$$\begin{aligned} x_f(t) &= U(t,s)u(s)\int_{t_0}^{s} \phi_n(\xi)\|u(\xi)\|^{-1}d\xi + \int_s^t U(t,\xi)u(\xi)\phi_n(\xi)\|u(\xi)\|^{-1}d\xi \\ &= U(t,s)x_f(s) + \int_s^t U(t,\xi)f(\xi)d\xi, \quad \forall\, t \ge s, \end{aligned}$$

i.e., $x_f \in C_0(\mathbb{R}, \mathbb{X})$ is a solution to Eq. (2.1). By assumption and (2.4) we have

$$\|x_f\| \le c\|f\|,$$

so, since $\|f\| = 1$,

$$\|u(t)\| \int_{t_0}^{t} \frac{\phi_n(s)}{\|u(s)\|}ds \le c, \forall t \ge t_0.$$

Letting $n \to \infty$ we get

$$\|u(t)\| \int_{t_0}^{t} \|u(s)\|^{-1}ds \le c, \forall t \ge t_0. \tag{2.7}$$

It is easy to see that the above inequality and the bounded growth of the process yield that there are positive constants K, β independent of x, t_0 such that

$$\|U(t,t_0)x\| = u(t) \le Ke^{-\beta(t-s)}u(s) = Ke^{-\beta(t-s)}U(s,t_0), \quad \forall t \ge s \ge t_0. \tag{2.8}$$

In fact, set $g(t) = \int_{t_0}^t \|u(s)\|^{-1}ds$. Then by (2.7) we obtain

$$\frac{d}{dt}\ln g(t) \ge \frac{1}{c}, \quad \forall t > t_0.$$

Next, assuming that $\tau > 1$ is given, integrating both sides of the above over the interval $[t_0+1, t_0+\tau)$ we get

$$\ln g(t_0+\tau) - \ln g(t_0+1) \ge \frac{\tau-1}{c}.$$

Hence,

$$g(t_0+\tau) \ge g(t_0+1)e^{\frac{\tau-1}{c}}. \tag{2.9}$$

Now, by the bounded growth, i.e., $\|U(t,s)\| \le Ne^{\omega(t-s)}$, $\forall t \ge s$, for any $s \in [t_0, t_0+1]$ we have

$$\|u(s)\| \le \|u(t_0)\|Ne^{\omega}.$$

Consequently, without loss of generality assuming that $t_1 > t_0+1$ we have

$$g(t_0+1) = \int_{t_0}^{t_0+1} \|u(s)\|^{-1}ds \ge \frac{1}{N}e^{-\omega}\|u(t_0)\|^{-1}. \tag{2.10}$$

By (2.7), for $\tau > 1$ we get

$$\|u(t_0+\tau)\| \le \frac{c}{g(t_0+\tau)}.$$

By (2.9),

$$\frac{c}{g(t_0+\tau)} \le \frac{ce^{-(\tau-1)c}}{g(t_0+1)}.$$

Next, by (2.10)

$$\frac{ce^{-(\tau-1)c}}{g(t_0+1)} \le Ne^{\omega}\|u(t_0)\|ce^{-(\tau-1)c} = N_1e^{-\tau/c}\|u(t_0)\|$$

where

$$N_1 = cNe^{1/c+\omega}.$$

Thus for $\tau > 1$ we have

$$\|u(t_0+\tau)\| \le N_1e^{-\tau/c}\|u(t_0)\| \tag{2.11}$$

On the other hand, for $0 \le \tau \le 1$ using the bounded growth we have

$$\|u(t_0+\tau)\| \le Ne^{1/c+\omega}e^{-\tau/c}\|u(t_0)\|. \tag{2.12}$$

Finally, combining (2.12) with (2.11) we have

$$\|u(t_0+\tau)\| \le N_2e^{-\tau/c}\|u(t_0)\|, \quad \forall 0 < \tau < t_1,$$

where $N_2 := \max\{N_1, Ne^{1/c+\omega}\}$. This finishes the proof. □

Corollary 2.1. *Let (ii) hold. Then*

$$X_1(t_0) := \{x \in \mathbb{X} : \lim_{t\to+\infty} U(t,t_0)x = 0\} \tag{2.13}$$

is a closed subspace of $\mathbb{X}$.

Proof. This corollary is an immediate consequence of the above lemma. □

To proceed we need a notion of extension.

Definition 2.3. We say that the function $u(t) := U(t,t_0)x$ for $t \geq t_0$ is extendable to a function on the whole line as a solution of the corresponding homogeneous equation (also denoted by $u(\cdot)$), if u is extendable to the whole line such that $u(t) = U(t,s)u(s)$, $\forall t \geq s, t, s \in \mathbb{R}$.

Note that if (ii) holds, then $u(t)$ has at most one extension on $(-\infty, t_0]$ such that it vanishes at infinity. In fact if we have two extensions, then by denoting by $u_1(t)$ and $u_2(t)$ the two functions defined on the whole line, respectively, we see that

$$w(t) = \begin{cases} 0, \forall t \geq t_0 \\ u_1(t) - u_2(t), \forall t \leq t_0 \end{cases}$$

is a solution in $C_0(\mathbb{R},\mathbb{X})$ to (2.1) with $f = 0$. Hence, by (ii), $u_1(t) = u_2(t)$, $\forall t \leq t_0$.

Lemma 2.3. *Let (ii) hold and let* $u(t) := U(t,t_0)x$. *Then there exist positive constants* K', β' *independent of* x *such that*

$$\|u(t)\| \leq K'e^{-\beta'(s-t)}\|u(s)\|, \quad \forall t \leq s \leq t_0, \tag{2.14}$$

provided that u *has an extension to the whole line as a solution of the corresponding homogeneous equation vanishing at* $-\infty$.

Proof. Suppose that $u(t) := U(t,t_0)x$, $t \geq t_0$ can be extended to the whole line as a solution to Eq. (2.1). Consider the function

$$f(t) = -\frac{\varphi_n(t)}{\|u(t)\|}u(t), \quad t \in \mathbb{R}, \tag{2.15}$$

where for every $n \in \mathbb{N}$

$$\varphi_n(t) := \begin{cases} 1, \;\; \forall 2/n \leq t \leq t_0 + n, \\ 1 - (t - t_0 - n), \;\; \forall t_0 + n \leq t \leq t_0 + n + 1 \\ 0, \;\; \forall t \geq t_0 + n + 1, \text{ or } t \leq 1/n. \end{cases}$$

Clearly that

$$y(t) = u(t) \int_t^\infty \varphi_n(s) \|u(s)\|^{-1} ds$$

is a solution of (2.1) with f defined by (2.15). By the same reasoning as above and by letting n to infinity we can show that

$$\|u(t)\| \int_t^\infty \|u(s)\|^{-1} ds \le c,$$

where c is defined by (2.4). Finally, as above we can prove (2.14). □

A consequence of this lemma is the following:

Corollary 2.2. *Let (ii) hold. Then*

$$X_2(t_0) := \{x \in \mathbb{X} : u(t) := U(t, t_0)x \text{ has an extension on } (-\infty, t_0] \text{ as a solution to the homogeneous equation vanishing at } -\infty\}$$

is a closed subspace of $\mathbb{X}$.

Lemma 2.4. *Let (ii) hold. Then for every* $t_0 \in \mathbb{R}$

$$\mathbb{X} = X_1(t_0) \oplus X_2(t_0), \tag{2.16}$$

where $X_1(t_0)$ *and* $X_2(t_0)$ *are defined by (2.13) and (2.16).*

Proof. For every $x \in \mathbb{X}$ let $\phi(t)$ be a continuously differentiable function on $\mathbb{R}$ with the following properties: $supp\phi \subset [t_0, +\infty)$, $\phi(t) = 1, \forall t \ge t_0 + 1$ and $|\phi'(t)| \le 2$. Next, let $f(t) = \phi'(t)U(t, t_0)x$. Obviously, f can be continuously extended to the whole line by setting $f(t) = 0, \forall t \le t_0$. By the assumption there exists a unique solution $x_f \in C_0(\mathbb{R}, \mathbb{X})$ to (2.1). The function $v(t) := \phi(t)U(t, t_0)x$, $t \ge t_0$ is also a solution on $[t_0, +\infty)$ of (2.1). Set $z(t) := v(t) - x_f(t)$ and $y(t) = U(t, t_0)x - z(t)$ for $t \ge t_0$. Then obviously $z(t)$ and $y(t)$ are solutions of the corresponding homogeneous equation of (2.1) on $(-\infty, +\infty)$ and $[t_0, \infty)$, respectively. By definition, we have

$$\lim_{t \to -\infty} \|z(t)\| = 0, \quad \lim_{t \to +\infty} \|y(t)\| = 0$$

that is, $z(t_0) = -x_f(t_0) \in \mathbb{X}_2(t_0)$ and $y(t_0) = x - z(t_0) \in \mathbb{X}_1(t_0)$. This proves the lemma. □

Now we finish the proof of the theorem by choosing the projection $Q(t)$ as the projection of $\mathbb{X} = \mathbb{X}_1(t) \oplus \mathbb{X}_2(t)$ onto $\mathbb{X}_2(t)$ for every $t \in \mathbb{R}$. □

Remark 2.1. If the evolutionary process $(U(t,s))_{t\geq s}$ is determined by a C_0-semigroup $T((t))_{t\geq 0}$ that is generated by an operator A, then the condition $\sigma(A) \cap i\mathbb{R} = \emptyset$ does not guarantee the hyperbolicity of $T((t))_{t\geq 0}$, so it does not imply the exponential dichotomy of $(U(t,s))_{t\geq s}$. This is because in general the Spectral Mapping Theorem does not hold for arbitrary C_0-semigroup. As a counterexample we can take the semigroup $(T(t))_{t\geq 0}$ in Example 1.12. Obviously, since $\sigma(A) = \emptyset$, we have $\sigma(A) \cap i\mathbb{R} = \emptyset$ and $\|T(t)\| = 1$ for all $t \geq 0$. Hence, if $(T(t))_{t\geq 0}$ is hyperbolic, then the projection Q in the Definition 2.2 must be trivial, i.e., $Q = I$. That means $\|T(t)\| \leq Ke^{-\alpha t}$ for all $t \geq 0$, where K, α are some positive numbers. But this contradicts to the fact that $\|T(t)\| = 1$ for all $t \geq 0$.

2.2 Evolution Semigroups and Perron Theorem

In this section we will discuss the relation between the Perron Theorem and the so-called evolution semigroup of bounded linear operators on $C_0(\mathbb{R}, \mathbb{X})$ associated with a given evolutionary process.

Definition 2.4. Let $(U(t,s))_{t\geq s}$ be a strongly continuous evolutionary process on $\mathbb{R}$ in a Banach space $\mathbb{X}$. Then the following family $(T^h)_{h\geq 0}$ of linear operators on $C_0(\mathbb{R}, \mathbb{X})$ is called the evolution semigroup associated with $(U(t,s))_{t\geq s}$ on the function space $C_0(\mathbb{R}, \mathbb{X})$:

$$\left[T^h v\right](t) = U(t, t-h)v(t-h), \quad \forall v \in C_0(\mathbb{R}, \mathbb{X}), h \geq 0, t \in \mathbb{R}. \tag{2.17}$$

Lemma 2.5. *$(T^h)_{h\geq 0}$ is a strongly continuous semigroup of linear operators on $C_0(\mathbb{R}, \mathbb{X})$ with the infinitesimal generator $\mathcal{G}$ such that $D(\mathcal{G}) = D(\mathcal{L})$ and $\mathcal{G}u = -\mathcal{L}u$ for all $u \in D(\mathcal{L})$.*

Proof. First, we show that for every $v \in C_0(\mathbb{R}, \mathbb{X})$ with compact support we have

$$\lim_{h\downarrow 0} T^h v = v.$$

Since *suppv* is compact, the range $R(v)$ of v is compact. By the strong continuity of the process $(U(t,s))_{t\geq s}$ we have that the map $\Delta \times \mathbb{X} \ni (t,s,x) \mapsto U(t,s)x \in \mathbb{X}$ is continuous (here $\Delta := \{(t,s) \in \mathbb{R}^2 : t \geq s\}$. Thus it is uniformly continuous on $(\Delta \cap [0,t_0]^2) \times R(v)$. On the other hand, v is uniformly continuous on $\mathbb{R}$. Thus, for every $\varepsilon > 0$ there exists a positive δ such that for $|h| < \delta$

$$\|T^h v - v\| = \sup_{t\in\mathbb{R}} \|U(t,t-h)v(t-h) - v(t)\| < \varepsilon.$$

Obviously, since there are positive constants M, ω such that $\|U(t,s)\| \geq Me^{\omega(t-s)}$ for all $t \geq s$, we have $\|T^h v\| \leq Me^h \|v\|$, $\forall h \geq 0$. Thus, for arbitrary $w \in C_0(\mathbb{R}, \mathbb{X})$ there is a sequence v_n with compact support in $C_0(\mathbb{R}, \mathbb{X}))$ convergent to w as $n \to \infty$. We have

$$0 \leq \limsup_{h \downarrow 0} \|T^h w - w\| \leq (1 + Me^{\omega})\|v_n - w\|.$$

This shows that $\lim_{h\downarrow 0} T^h w = w$, i.e., the strong continuity of the evolution semigroup.

Now we suppose that $u \in D(\mathcal{L})$. Then there is a unique $f \in C_0(\mathbb{R}, \mathbb{X})$ such that (2.1) holds. We have

$$\begin{aligned}\left[\frac{T^h u - u}{h}\right](t) &= \frac{1}{h}\left(U(t, t-h)u(t-h) - u(t)\right) \\ &= -\frac{1}{h}\int_{t-h}^{t} U(t,\xi)f(\xi)d\xi.\end{aligned}$$

Since $f \in C_0(\mathbb{R}, \mathbb{X})$, f is uniformly continuous and $R(f)$ is relatively compact. Consequently, $(t, s, x) \mapsto U(t, s)x$ is uniformly continuous on $(\Delta \cap [0, t_0]^2) \times R(f)$. This yields that

$$\lim_{h\downarrow 0} \sup_{t \in \mathbb{R}} \frac{1}{h}\int_{t-h}^{t} \|U(t,\xi)f(\xi) - f(\xi)\| d\xi = 0,$$

i.e.,

$$\lim_{h\downarrow 0}\left[\frac{T^h u - u}{h}\right] = -f. \tag{2.18}$$

Conversely, let $u \in D(\mathcal{G})$. By the basic properties of C_0-semigroups we have that the map $(0, \infty) \ni h \mapsto T^h u$ is differentiable and

$$\frac{d}{dh}T^h u = \mathcal{G}u, \ \forall h > 0$$

and

$$T^h u - u = \int_0^h T^{\xi}\mathcal{G}u d\xi, \ \forall h \geq 0.$$

Setting $f := \mathcal{G}u$ we have

$$\begin{aligned}\left[T^h u - u\right](t) &= U(t, t-h)u(t-h) - u(t) = \int_0^h \left[[T\xi f\right](t) d\xi \\ &= \int_{t-h}^{t} U(t,\xi)f(\xi)d\xi.\end{aligned}$$

Hence,

$$u(t) = U(t, t-h)u(t-h) - \int_{t-h}^{t} U(t,\xi)f(\xi)d\xi, \quad \forall t \in \mathbb{R}, h \geq 0.$$

This shows that u is a solution of (2.1) with the forcing term $-f$ and finishes the proof of the lemma. □

Remark 2.2. It is easy to see that if

$$U(t,s) = e^{(t-s)A} := \sum_{k=0}^{\infty} \frac{((t-s)A)^k}{k!},$$

where $A \in L(\mathbb{X})$, then $D(\mathcal{G}) = \{u \in C_0(\mathbb{R}, \mathbb{X}) : \exists u' \in C_0(\mathbb{R}, \mathbb{X})\}$ and $\mathcal{G}$ is of the form

$$\mathcal{G}u(t) = -\frac{d}{dt}u(t) + Au(t), \quad \forall u \in D(\mathcal{G}).$$

Let $U(t,s) = T(t-s)$ for certain C_0-semigroup $(T(t))_{t\geq 0}$ with generator A. For unbounded A, the above formula is no longer true. To see how it should be modified we consider the evolution semigroup $(T^h)_{h\geq 0}$ as the product of two commutative semigroups $(R(h))_{h\geq 0}$ and $(S(h))_{h\geq 0}$ defined as follows

$$[R(h)v](t) = T(h)v(t); \quad [S(h)v](t) = v(t-h), \quad \forall h \geq 0, v \in C_0(\mathbb{R}, \mathbb{X}).$$

Let $\mathcal{A}$ be the operator with

$$D(\mathcal{A}) = \{u \in C_0(\mathbb{R}, \mathbb{X}) : u(t) \in D(A), \ \forall t \in \mathbb{R}, \ Au(\cdot) \in C_0(\mathbb{R}, \mathbb{X})\},$$

and $\mathcal{A}u = Au(t)$, $\forall t \in \mathbb{R}, u \in D(\mathcal{A})$. Clearly that the infinitesimal generator of $(S(t))_{t\in\mathbb{R}}$ is $-d/dt$ with $D(-d/dt) = \{f \in C_0(\mathbb{R}, \mathbb{X}) : \exists f' \in C_0(\mathbb{R}, \mathbb{X})\}$. Thus, by Exercise 7, we have

$$\mathcal{G} = \overline{-d/dt + \mathcal{A}}.$$

As an immediate consequence of the above lemma we have

Corollary 2.3. *The process $(U(t,s))_{t\geq s}$ has an exponential dichotomy if and only if the generator $\mathcal{G}$ of the evolution semigroup $(T^h)_{h\geq 0}$ associated with $(U(t,s))_{t\geq s}$ is invertible.*

Corollary 2.4. *If $\lambda \in \sigma(\mathcal{G})$, then $\lambda + i\mu \in \sigma(\mathcal{G})$ for all $\mu \in \mathbb{R}$.*

Proof. Let $U_\mu(t,s) := e^{-i\mu(t-s)}U(t,s)$, for fixed $\mu \in \mathbb{R}$. Note that $(U(t,s))_{t\geq s}$ has an exponential dichotomy if and only if so does $(U_\mu(t,s))_{t\geq s}$. Obviously, the generator of the evolution semigroup associated with $(U_\mu(t,s))_{t\geq s}$ is the operator $\mathcal{G} - i\mu I$ with the same domain. By the above corollary, $\mathcal{G}$ is invertible if and only if so is $\mathcal{G} - i\mu I$. □

Definition 2.5. Let $(T(t))_{t\geq 0}$ be a C_0-semigroup on $\mathbb{X}$. It is said to be hyperbolic if $\sigma(T(1)) \cap \Gamma = \emptyset$, where Γ denotes the unit circle on the complex plane.

Theorem 2.2. *Let $(U(t,s))_{t\geq s}$ be a strongly continuous evolutionary process and let $(T^h)_{h\geq 0}$ be its evolution semigroup associated with it on $C_0(\mathbb{R},\mathbb{X})$. Then $(U(t,s))_{t\geq s}$ has an exponential dichotomy if and only if $(T^h)_{h\geq 0}$ is hyperbolic.*

Proof. By the inclusion of C_0-semigroups, if $\sigma(T^1)\cap\Gamma=\emptyset$, then $0\notin\sigma(\mathcal{G})$. By the above corollaries $(U(t,s))_{t\geq s}$ has an exponential dichotomy. Conversely, if $(U(t,s))_{t\geq s}$ has an exponential dichotomy, then we set $C_1:=\{u\in C_0(\mathbb{R},\mathbb{X}): u(t)\in ImP(t),\ \forall t\in\mathbb{R}\}$ and $C_2:=\{u\in C_0(\mathbb{R},\mathbb{X}): u(t)\in ImQ(t),\ \forall t\in\mathbb{R}\}$, where projections $P(t),Q(t),\ t\in\mathbb{R}$ are determined from the exponential dichotomy of $(U(t,s))_{t\geq s}$. It is easy to see that T^1 is strictly contractive on C_1 and is left C_2 invariant. Moreover, the restriction of T^1 to C_2 is invertible and its inverse is also strictly contractive. By the Spectral Radius Theorem $\sigma(T^1)\cap\Gamma=\emptyset$. □

The following spectral mapping theorem holds for evolution semigroups.

Theorem 2.3. *We have the spectral mapping theorem for the evolution semigroup $(T^h)_{h\geq 0}$*

$$\sigma(T^h)\backslash\{0\}=e^{h\sigma(\mathcal{G})},\quad \forall h\geq 0. \tag{2.19}$$

Proof. By the above theorem $1\in\rho(T^1)$ if and only if $\Gamma\subset\rho(T^1)$. For any positive $\mu\in\mathbb{R}_+$ consider the process $V_\mu(t,s):=e^{-\mu(t-s)}U(t,s)$, applying the above theorem to this process we have that $\mu\in\rho(T^1)$ if and only if the whole circle $\{|z|=\mu\}$ is contained in $\rho(T^1)$. Since $0\in\rho(\mathcal{G})$ if and only if $1\in\rho(T^1)$, by Lemma 2.4 we have

$$\sigma(T^1)\backslash\{0\}=e^{\sigma(\mathcal{G})}.$$

For arbitrary $\eta>0$ we consider the process $V_h(t,s):=e^{-\eta(t-s)}U(t,s)$. Its evolution semigroup is $T^h_\eta)_{h\geq 0}$ defined by $T^h_\eta v(t)=e^{-\eta}T^h$. Hence, $1\in\rho(T^1_\eta)=e^\eta\rho(T^1)$ if and only if $\Gamma\subset\rho(T^1_\eta)=e^\eta\rho(T^1)$. Repeating the above reasoning we have (2.19). □

Next, we consider the perturbation theory of exponential dichotomy of processes. Let $(U(t,s))_{t\geq s}$ be a strongly continuous evolutionary process and let $B:\mathbb{R}\to L(\mathbb{X})$ be continuous. Then we can associated with the following integral equation

$$x(t)=U(t,s)x(s)+\int_s^t U(t,\xi)B(\xi)x(\xi)d\xi,\quad \forall t\geq s \tag{2.20}$$

a strongly continuous evolutionary process $(V(t,s))_{t\geq s}$ as follows: for any $t\geq s$ we define $V(t,s)y$ as the unique solution to the above equation such that $x(s) = y$. We can show without difficulty that such a family of operators forms a strongly continuous evolutionary process. Moreover, denoting its evolutionary process by $(T_B^h)_{t\geq s}$ we can prove that $\|T_B^1 - T^1\| \leq \sup_{t\in\mathbb{R}} \|B(t)\|$. Hence, we have the following

Corollary 2.5. *Let $(U(t,s))_{t\geq s}$ have an exponential dichotomy. Then the process $(V(t,s))_{t\geq s}$ defined as above has an exponential dichotomy as well provided that $\sup_{t\in\mathbb{R}} \|B(t)\|$ is sufficiently small.*

Proof. By the above theory, $1 \in \rho(T^1)$. For sufficiently small $\sup_{t\in\mathbb{R}} \|B(t)\|$ we have $1 \in \rho(T_B^1)$. This yields the hyperbolicity of T_B^h, and hence the exponential dichotomy of $(V(t,s))_{t\geq s}$ follows. □

Example 2.1. If $U(t,s) = T(t-s)$, where $(T(t))_{t\geq 0}$ is a C_0-semigroup, then we can easily show that $(U(t,s))_{t\geq s}$ has an exponential dichotomy if and only if $T(1)$ is hyperbolic. In fact, in this case we consider the Riesz projection

$$P = \frac{1}{2\pi i}\int_\Gamma (\lambda - T(1))^{-1} d\lambda,$$

where Γ is positively oriented. By the general theory of linear operators $\mathbb{X} = ImP \oplus KerP$ and, $\sigma(T(1)) = \sigma(T(1)|_{ImP}) \sqcup \sigma(T(1)|_{KerP})$. Hence, by the Spectral Radius Theorem there are positive constants K, α such that

$$\|T(t)x\| \leq Ke^{-\alpha t}\|x\|, \quad \forall t \geq 0; x \in ImP$$
$$\|T(s)y\| \leq Ke^{-\alpha s}\|y\|, \quad \forall s \leq 0; y \in KerP,$$

where $T(s)|_{KerP}$ is the inverse of $T(-s)|_{KerP}$ for $s > 0$ whose existence is guaranteed by the above decomposition of spectrum of $T(1)$. Thus, we can check that the process $U(t,s) = T(t-s)$ has an exponential dichotomy.

If $(T(t))_{t\geq 0}$ is eventually norm continuous with generator A, then by Spectral Mapping Theorem, for $(T(t))_{t\geq 0}$ to have an exponential dichotomy it is necessary and sufficient that

$$\sigma(A) \cap i\mathbb{R} = \emptyset.$$

2.3 Stability Theory

2.3.1 *Exponential Stability*

In this subsection we consider the exponential stability of a strongly continuous evolutionary process $(U(t,s))_{t\geq s}$ with $t, s \in \mathbb{R}_+$.

Definition 2.6. An evolutionary process $(U(t,s))_{t\geq s}$ is said to be exponentially stable if there are positive constants K, α such that

$$\|U(t,s)\| \leq Ke^{-\alpha(t-s)}, \quad \forall t \geq s.$$

The main result of this subsection is the following

Theorem 2.4. *Let $(U(t,s))_{t\geq s}$ with $t \geq s \geq 0$ be a strongly continuous evolutionary process. Then, $(U(t,s))_{t\geq s}$ is exponentially stable if and only if for every $f \in C_0(\mathbb{R}, \mathbb{X})$ the following function*

$$u_f(t) = \int_0^t U(t,\xi)f(\xi)d\xi, \quad t \in \mathbb{R}_+ \tag{2.21}$$

is in $C_0(\mathbb{R}, \mathbb{X})$.

Proof. For every $t \geq 0$ we define an operator $V(t) \in L(C_0(\mathbb{R}, \mathbb{X}), \mathbb{X})$

$$V(t) : C_0(\mathbb{R}, \mathbb{X}) \ni f \mapsto \int_0^t U(t,\xi)f(\xi)d\xi \in \mathbb{X}.$$

By the assumption, for every $f \in C_0(\mathbb{R}, \mathbb{X})$ we have

$$\sup_{t\geq 0} \|V(t)f\| < \infty.$$

Thus, by the Uniform Boundedness Principle,

$$\sup_{t\geq 0} \|V(t)\| < \infty.$$

Consequently, there is a positive constant $c > 0$ such that

$$\|u_f\| \leq c\|f\|, \quad f \in C_0(\mathbb{R}, \mathbb{X}).$$

By the same reasoning as in the proof of Lemma 2.2 we can show that there are positive constants K, α depending only on c such that

$$\|U(t,s)\| \leq Ke^{-\alpha(t-s)}, \quad \forall t \geq s \geq 0.$$

This finishes the proof of the theorem. □

We are now interested in the stability of individual orbits in the case the process is not exponentially stable. To this end we introduce an operator $\mathcal{G} : D(\mathcal{G}) \subset BC(\mathbb{R}_+, \mathbb{X}) \to BC(\mathbb{R}_+, \mathbb{X})$, where we denote by $BC(\mathbb{R}_+, \mathbb{X})$ the space of all $\mathbb{X}$ valued bounded and continuous functions on $\mathbb{R}_+$ with sup-norm as follows:

$$D(\mathcal{G}) := \{u \in BC(\mathbb{R}_+, \mathbb{X}) \text{ such that } \exists f \in BC(\mathbb{R}_+, \mathbb{X})$$
$$u(t) = \int_0^t U(t\xi)f(\xi)d\xi, \ \forall t \in \mathbb{R}_+\}$$

$$\mathcal{G}u := f, \quad \forall u \in D(\mathcal{G}).$$

As for the operator $\mathcal{L}$ in the previous section we can show that $\mathcal{G}$ is closed.

Theorem 2.5. *Assume that there is a positive constant $\nu > 0$ such that*

$$\|\mathcal{G}u\| \geq \nu\|u\|, \quad \forall u \in D(\mathcal{G}). \tag{2.22}$$

Then there exist positive constants K, α depending only on ν such that

$$\|U(t,t_0)x\| \leq Ke^{-\alpha(t-s)}\|U(s,t_0)x\|, \quad \forall t \geq s \geq t_0$$

provided that

$$\sup_{t\geq 0}\|U(t,t_0)x\| < \infty.$$

Proof. The proof can be done in the same manner as in Lemma 2.2, so the details are omitted. □

Now we consider the perturbation theory of an exponentially stable process $(U(t,s))_{t\geq s}$. As in the previous section for a given bounded $B : \mathbb{R}_+ \to L(\mathbb{X})$ we can show that the following integral equation

$$x(t) = U(t,s)x(s) + \int_s^t U(t,\xi)B(\xi)x(\xi)d\xi, \quad \forall t \geq s \geq 0$$

generates a strongly continuous evolutionary process $(V(t,s))_{t\geq s}$. Using the Gronwall inequality it is easy to prove the following assertion:

Proposition 2.1. *Let $(U(t,s))_{t\geq s}$ be an exponentially stable strongly continuous evolutionary process. Then the process $(V(t,s))_{t\geq s}$ is also exponentially stable provided that $\sup_{t\geq 0}\|B(t)\|$ is sufficiently small.*

Proof. Let

$$\varepsilon := \sup_{t\geq 0}\|B(t)\|$$

and $\eta(t) := V(t,s)x$ for any fixed $s \in \mathbb{R}_+, x \in \mathbb{X}$. Then

$$\eta(t) \leq Ke^{-\alpha(t-s)}\|x\| + \int_s^t \varepsilon Ke^{-\alpha(t-\xi)}\eta(\xi)d\xi, \quad \forall t \geq s.$$

Applying the Gronwall inequality to the function $\zeta(t) := e^{\alpha t}\eta(t)$ we get

$$\eta(t) \leq Ke^{-(\alpha-K\varepsilon)(t-s)}\|x\|, \quad \forall t \geq s.$$

Thus

$$\|V(t,s)\| \leq Ke^{-(\alpha-K\varepsilon)(t-s)}, \quad \forall t \geq s.$$

Therefore, if

$$\varepsilon < \frac{\alpha}{K}$$

then $(V(t,s))_{t\geq s}$ is exponentially stable. □

2.3.2 Strong Stability

In this subsection we will prove the following theorem: if $(T(t))_{t\geq 0}$ is a uniformly bounded C_0-semigroup such that the unitary spectrum $\sigma(A)\cap i\mathbb{R}$ is at most countable and $\sigma_p(A^*)\cap i\mathbb{R} = \emptyset$, then $(T(t))_{t\geq 0}$ is strongly stable, i.e. $\lim_{t\to\infty} \|T(t)x\| = 0$ for all $x \in \mathbb{X}$.

Definition 2.7. A bounded operator T on a Banach space $\mathbb{X}$ is called an *isometry* if $\|Tx\| = \|x\|$ for all $x \in \mathbb{X}$.

Lemma 2.6. *Let $(T(t))_{t\geq 0}$ be a C_0-semigroup of isometries on a Banach space $\mathbb{X}$, and let A be its generator. Then the following hold:*

(i) For all $x \in D(A)$ and $\lambda \in \mathbb{C}$ we have $\|(\lambda - A)x\| \geq |\Re\lambda|\, \|x\|$;
(ii) If $E \subset \mathbb{R}$ is closed and $x \in \mathbb{X}$ is such that the map $\lambda \mapsto R(\lambda, A)x$ has a holomorphic extension F to a connected neighbourhood V of $\{\Re\lambda \geq 0\}\backslash iE$, then for all $\lambda \in V\backslash i\mathbb{R}$ we have $\|F(\lambda)\| \leq |\Re\lambda|^{-1}\|x\|$.

Proof. (i): First, we may assume that $\Re\lambda \neq 0$. From the identity

$$e^{-\lambda t}T(t)x = x + \int_0^t e^{-\lambda s}(A-\lambda)T(s)x\,ds$$

we have

$$\begin{aligned} e^{-\mathrm{Re}\,\lambda t}\|x\| &= e^{-\mathrm{Re}\,\lambda t}\|T(t)x\| \\ &\leq \|x\| + \int_0^t e^{-\mathrm{Re}\,\lambda s}\|T(s)(\lambda - A)x\|\,ds \\ &= \|x\| + \left(\int_0^t e^{-\mathrm{Re}\,\lambda s}\,ds\right)\|(\lambda - A)x\| \\ &= \|x\| + \frac{e^{-\mathrm{Re}\,\lambda t} - 1}{-\mathrm{Re}\,\lambda}\|(\lambda - A)x\|. \end{aligned}$$

This proves the lemma for $\Re\lambda < 0$. For $\Re\lambda > 0$ the inequality follows from the Laplace transform representation of the resolvent.

(ii): This is proved in the same way, after first substituting $R(\lambda, A)x$ for x in the first formula and passing to the holomorphic extension. □

Theorem 2.6. *Let $(T(t))_{t\geq 0}$ be C_0-semigroup of contractions on a Banach space X, with generator A. Then there exists a Banach space Y, a bounded operator $\pi : X \to Y$ with dense range, and a C_0-semigroup $(U(t))_{t\geq 0}$ of isometries on Y with generator B such that:*

(i) $U(t)\pi = \pi T(t)$ for all $t \geq 0$. Moreover, $\pi D(A) \subset D(B)$ and $B\pi x = \pi Ax$ for all $x \in D(A)$;

(ii) $\lim_{t\to\infty} \|T(t)x\| = \|\pi x\|$ *for all* $x \in X$*;*
(iii) $\sigma(B) \subset \sigma(A)$.

If $\sigma(A) \cap i\mathbb{R}$ *is a proper subset of* $i\mathbb{R}$*, then* $(U(t))_{t\geq 0}$ *extends to a* C_0*-group of isometries.*

Proof. On X we define the seminorm l by $l(x) := \lim_{t\to\infty} \|T(t)x\|$. Since $(T(t))_{t\geq 0}$ is contractive, this limit indeed exists. Let $\pi : X \to Y_0 := X/\ker l$ be the quotient mapping. The seminorm l induces a norm l_0 on Y_0 by $l_0(\pi x) := l(x)$, and hence

$$l_0(\pi x) = l(x) = \lim_{t\to\infty} \|T(t)x\|.$$

For $t \geq 0$, we define $U_0(t) : Y_0 \to Y_0$ by $U_0(t)\pi x := \pi T(t)x$. We have

$$l_0(U_0(t)\pi x) = l_0(\pi T(t)x) = l(T(t)x) = l(x) = l_0(\pi x).$$

This shows that $U_0(t)$ is isometric with respect to the norm l_0. Let Y be the completion of Y_0 with respect to l_0. Then each operator $U_0(t)$ extends to an isometry $U(t)$ on Y. Strong continuity of the family $(U(t))_{t\geq 0}$ follows from the density of πX in Y, the contractivity of the operators $U(t)$, and the estimate

$$\begin{aligned}\limsup_{t\downarrow 0} \|U(t)\pi x - \pi x\| &= \limsup_{t\downarrow 0} \left(\lim_{s\to\infty} \|T(t+s)x - T(s)x\| \right) \\ &\leq \limsup_{t\downarrow 0} \|T(t)x - x\| = 0, \quad x \in X.\end{aligned}$$

If $x \in D(A)$, then

$$\lim_{t\downarrow 0} \frac{1}{t}(U(t)\pi x - \pi x) = \pi \lim_{t\downarrow 0} \frac{1}{t}(T(t)x - x) = \pi Ax,$$

proving that $\pi x \in D(B)$ and $B\pi x = \pi Ax$.

We have proved (i) and (ii). Next, we prove (iii). Let $\lambda \in \varrho(A)$. We define a linear operator R_λ on Y_0 as follows

$$R_\lambda \pi x := \pi R(\lambda, A)x.$$

This operator is well-defined and

$$\begin{aligned}l_0(R_\lambda \pi x) &= \lim_{t\to\infty} \|T(t)R(\lambda, A)x\| \\ &\leq \|R(\lambda, A)\| \lim_{t\to\infty} \|T(t)x\| = \|\mathbb{R}\|\, l_0(\pi x).\end{aligned}$$

Therefore, R_λ extends to a bounded operator on Y and $\|R_\lambda\| \leq \|R(\lambda, A)\|$. For all $x \in X$ we have $R_\lambda \pi x = \pi R(\lambda, A)x \in \pi D(A) \subset D(B)$ and $(\lambda -$

$B)R_\lambda \pi x = (\lambda - B)\pi R(\lambda, A)x = \pi(\lambda - A)\mathbb{R}x = x$. Similarly, for all $x \in D(A)$ we have $\pi x \in D(B)$ and $R_\lambda(\lambda - B)\pi x = R_\lambda \pi(\lambda - A)x = \pi\mathbb{R}(\lambda - A)x = x$. Therefore, to prove that R_λ is a two-sided inverse of $\lambda - B$, in view of the closedness of B it remains to prove that $\pi D(A)$ is dense in $D(B)$ with respect to the graph norm. So let $y \in D(B)$ be arbitrary. Fix $\mu > 0$ arbitrary and choose $z \in Y$ such that $y = R(\mu, B)z$. Take a sequence $\{x_n\} \subset X$ such that $\pi x_n \to z$ in Y and put $y_n := \pi R(\mu, A)x_n$. Then $y_n \in \pi D(A)$, $y_n = R(\mu, B)\pi x_n \to R(\mu, B)z = y$, and $By_n = BR(\mu, B)\pi x_n \to BR(\mu, B)z = By$. Here we used that $\pi R(\mu, A) = R(\mu, B)\pi$ by the Laplace transform representation of the resolvents and that $BR(\mu, B) = \mu R(\mu, B) - I$ is a bounded operator on Y. Thus, $y_n \to y$ in $D(B)$ with respect to the graph norm. This concludes the proof of (iii).

Suppose $\sigma(A) \cap i\mathbb{R}$ is properly contained in $i\mathbb{R}$. We have to prove that $(U(t))_{t\geq 0}$ extends to a C_0-group of isometries. By Lemma 2.6 (i), for all $y \in D(B)$ and $\Re\lambda < 0$ we have

$$\|(\lambda - B)y\| \geq |\Re\lambda| \, \|y\|.$$

It follows that the open left half-plane $\mathbb{C}_-$ contains no approximate eigenvalues for B. In particular, $\mathbb{C}_-$ contains no elements of the boundary of $\sigma(B)$. Hence, either $\mathbb{C}_- \subset \sigma(B)$ or $\mathbb{C}_- \cap \sigma(B) = \emptyset$. But in the first case, also $i\mathbb{R} \subset \sigma(B)$ since $\sigma(B)$ is closed. This contradicts the assumption, so we must have the second alternative.

It follows that $\sigma(B) \subset i\mathbb{R}$. For $\Re\lambda > 0$ we have

$$\|R(\lambda, -B)\| = \|R(-\lambda, B)\| \leq \frac{1}{|\Re(-\lambda)|} = \frac{1}{\Re\lambda}.$$

By the Hille-Yosida theorem, $-B$ is the generator of a C_0-semigroup $(V(t))_{t\geq 0}$ of contractions on Y. We check that $U(t)$ is invertible for all $t \geq 0$ with inverse $V(t)$. For all $x \in D(B) = D(-B)$, the maps $t \mapsto U(t)V(t)x$ and $t \mapsto V(t)U(t)x$ are differentiable with derivative identically zero. It follows that the maps are constant, and by letting $t \downarrow 0$ it follows that $U(t)V(t) = V(t)U(t) = I$ on the dense set $D(B)$, hence on all of Y. Finally, each operator $V(t)$ is an isometry, being the inverse of an isometry. □

The triple $(Y, \pi, (U(t))_{t\geq 0})$ will be called the *isometric limit (semi)group* associated to $(T(t))_{t\geq 0}$.

Corollary 2.6. *If $(T(t))_{t\geq 0}$ is an isometric C_0-semigroup on $\mathbb{X}$ with $\sigma(A) \cap i\mathbb{R} \neq i\mathbb{R}$, then $(T(t))_{t\geq 0}$ extends to an isometric C_0-group.*

Proof. By the isometric nature of $(T(t))_{t\geq 0}$ we have $l(x) = x$ for all $x \in X$, so $Y = X$ and $(T(t))_{t\geq 0} = (U(t))_{t\geq 0}$. But $(U(t))_{t\geq 0}$ extends to a C_0-group since $\sigma(A) \cap i\mathbb{R} \neq i\mathbb{R}$. □

An invertible operator is called *doubly power bounded* if

$$\sup_{k\in\mathbb{Z}} \|T^k\| < \infty.$$

Lemma 2.7. *Let T be a doubly power bounded operator on a Banach space X with $\sigma(T) = \{1\}$. Then $T = I$.*

Proof. Since $\ln z$ is holomorphic in a neighbourhood of $z = 1$, by Dunford calculus we may define the bounded operator $S := -i \ln T$. Then $T = e^{iS}$ and the spectral mapping theorem implies that $\sigma(mS) = \{0\}$ for all $m \in \mathbb{N}$. Also, for all $m \in \mathbb{N}$ we have $\sigma(\sin(mS)) = \sin(\sigma(mS)) = \{\sin 0\} = \{0\}$, and

$$\|(\sin(mS))^n\| = \left\| \left(\frac{T^m - T^{-m}}{2i} \right)^n \right\| \leq \sup_{k\in\mathbb{Z}} \|T^k\|.$$

Let $\sum_{n=0}^{\infty} c_n z^n$ be Taylor series of the principle branch of $\arcsin z$ at $z = 0$. As is well-known, $c_n \geq 0$ for all n and $\sum_{n=0}^{\infty} c_n = \arcsin(1) = \frac{\pi}{2}$. Consequently,

$$\|mS\| = \|\arcsin(\sin(mS))\| \leq \sum_{n=0}^{\infty} c_n \|(\sin(mS))^n\| \leq \frac{\pi}{2} \sup_{k\in\mathbb{Z}} \|T^k\|.$$

Since this holds for all $m \in \mathbb{N}$, it follows that $S = 0$ and $T = e^{iS} = I$. □

Theorem 2.7. *Let $(T(t))_{t\geq 0}$ be a uniformly bounded C_0-semigroup on a Banach space $\mathbb{X}$, with generator A. If*

(i) $\sigma(A) \cap i\mathbb{R}$ is countable, and
(ii) $\sigma_p(A^*) \cap i\mathbb{R} = \emptyset$,

then $(T(t))_{t\geq 0}$ is strongly stable, i.e. $\lim_{t\to\infty} \|T(t)x\| = 0$ for all $x \in \mathbb{X}$.

Proof. By renorming with the equivalent norm $\|x\|_* := \sup_{t\geq 0} \|T(t)x\|$, we may assume that $(T(t))_{t\geq 0}$ is contractive. Let $(Y, \pi, (U(t))_{t\geq 0})$ be the isometric limit semigroup associated to $(T(t))_{t\geq 0}$, and let B be the generator of $(U(t))_{t\geq 0}$. By (i), $\sigma(A) \cap i\mathbb{R}$ cannot be all of $i\mathbb{R}$, and therefore $(U(t))_{t\geq 0}$ extends to an isometric group on Y. Assuming that $(T(t))_{t\geq 0}$ is not strongly stable, we shall prove that (i) implies $\sigma_p(A^*) \cap i\mathbb{R} \neq \emptyset$. In fact, since $(T(t))_{t\geq 0}$ is not strongly stable, the definition of Y implies that $Y \neq \{0\}$. By Lemma 1.1, $\sigma(B) \neq \emptyset$. Also, since $\sigma(B) \subset \sigma(A)$, it follows that $\sigma(B)$ is countable. In particular, since $\sigma(B)$ is closed and countable, it contains an isolated point, say $i\omega$. Let P_ω be the associated spectral projection in Y, let $(U_\omega(t))_{t\geq 0}$ be the restriction of $(U(t))_{t\geq 0}$ to $P_\omega Y$ and let B_ω denote its generator. Since $\sigma(B_\omega) = \{i\omega\}$, Theorem 1.10 implies that $\sigma(U_\omega(t)) = \{e^{i\omega t}\}$ for all $t \in \mathbb{R}$. By Lemma 2.7, this implies that $U_\omega(t) = e^{i\omega t}I$. Hence, for all $y \in Y$ we have $U_\omega(t)P_\omega y = e^{i\omega t}P_\omega y$, so $P_\omega y \in D(B_\omega)$ and $B_\omega P_\omega y = i\omega P_\omega y$. Fix an arbitrary non-zero $y^*_\omega \in (P_\omega Y)^*$ and define $x^* \in \mathbb{X}^*$ by

$$< x^*, x >:=< y^*_\omega, P_\omega \pi x >, \quad x \in \mathbb{X}.$$

Then $x^* \neq 0$. For all $x \in D(A)$ we have $\pi x \in D(B)$, $B\pi x = \pi A x$, and

$$\begin{aligned} < x^*, Ax > &= < y^*_\omega, P_\omega \pi A x >=< y^*_\omega, P_\omega B \pi x > \\ &= < y^*_\omega, B_\omega P_\omega \pi x >= i\omega < y^*_\omega, P_\omega \pi x >= i\omega < x^*, x > . \end{aligned}$$

Hence, $x^* \in D(A^*)$ and $A^* x^* = i\omega x^*$. □

2.4 Comments and Further Reading Guide

2.4.1 *Further Reading Guide*

So far we have discussed conditions for the stability and exponential dichotomy of homogeneous equations. Below we state without proofs further important results on the asymptotic behavior of C_0-semigroups. The reader is referred to Chap. 5 §7 in [van Neerven (78)]) for more information.

Definition 2.8. A C_0-semigroup $(T(t))_{t\geq 0}$ on $\mathbb{X}$ is said to be *almost periodic* if for each $x \in \mathbb{X}$ the set $\{T(t)x, t \in [0, +\infty)\}$ is relatively compact in $\mathbb{X}$.

Theorem 2.8. *(Theorem 5.7.10 in [van Neerven (78)]) Let $(T(t))_{t\geq 0}$ be a uniformly bounded C_0-semigroup on a Banach space $\mathbb{X}$, with generator A, and assume that $\sigma(A) \cap i\mathbb{R}$ is countable. Then the following assertions are equivalent:*

(1) $(T(t))_{t\geq 0}$ is almost periodic,
(2) For every $i\omega \in \sigma(A) \cap i\mathbb{R}$ the limit $\lim_{t\to\infty} \frac{1}{t}\int_0^t e^{-i\omega s}T(s)x\,ds$ exists for every $x \in \mathbb{X}$,
(3) For every $i\omega \in \sigma(A) \cap i\mathbb{R}$, $R(A - i\omega) + N(A - i\omega)$ is dense in $\mathbb{X}$.

The following is referred to as the splitting Theorem of Glicksberg and DeLeeuw.

Theorem 2.9. *(Theorem 5.7.7 in [van Neerven (78)]) Let $(T(t))_{t\geq 0}$ be an almost periodic C_0-semigroup on a Banach space $\mathbb{X}$. Then there exists a direct sum decomposition $\mathbb{X} = X_0 \oplus X_1$ of $(T(t))_{t\geq 0}$-invariant subspaces, where*

$$X_0 = \{x \in X : \lim_{t\to\infty} \|T(t)x\| = 0\}$$

and X_1 is the closed linear span of all eigenvectors of the generator A with purely imaginary eigenvalues. Moreover, the restriction of $(T(t))_{t\geq 0}$ to X_1 extends to an almost periodic C_0-group on X_1. If $(T(t))_{t\geq 0}$ is contractive, this group is isometric.

Exponential dichotomy of homogeneous equations that are defined on the half line is more difficult for us to study. Evolution semigroups method actually fails. However, Perron type characterization of exponential dichotomy can be used for these equations. We refer the reader to [Minh, Räbiger and Schnaubelt (70)] for more details in this direction. Nevertheless, evolution semigroups method is a strong tool to study the exponential dichotomy of linear skew products. We refer the reader to [Chicone and Latushkin (19)] for more details in this direction. New applications of evolution semigroups to the stability problem of homogeneous equations can be found in [Batty, Chill and Tomilov (15); Latushkin and Tomilov (56)]. The explicit formula for the generators of evolution semigroups gives rise to new applications of this method to the study of inhomogeneous linear and semilinear equations. This is the content of the next chapter.

2.4.2 *Comments*

Perron Theorem for evolutionary processes in Banach spaces was first proved by Zhikov (see [Zikov (108)]). The proof that is given in Section 1 is an adaptation of this proof for the function space $C_0(\mathbb{R}, \mathbb{X})$. Another recent proof of the characterization of exponential dichotomy using evolution semigroups was given by Latushkin and Montgomery-Smith in [Latushkin, Monthomery-Smith (55)] (see also [Chicone and Latushkin (19); van Neerven (78)]). The explicit formula for the generators of evolution semigroups was found in [Aulbach and Minh (10)] for general semilinear evolution equations. In the linear case, it was also found in [Baskakov (13)]. The presentation of Section 2 follows [Minh (69)].

The Perron condition (that is similar to the condition in Theorem 2.4) for exponential stability was first proved by Datko [Datko (25)]. The proof of Theorem 2.4 is taken from [Minh, Räbiger and Schnaubelt (70)]. Theorem 2.7, that is widely referred to as ABLV Theorem, was first proved by Sklyar and Shirman for the bounded case [Skylar and Shirman (97)] (we thank G.M. Sklyar for sending us their original paper). It is amazing that their result and method of proving can be extended to the unbounded case by Lyubich and Vu in [Lyubich and Vu (66)]. The ABLV Theorem was proved independently by Arendt and Batty in [Arendt and Batty (4)].

Chapter 3

Existence of Almost Periodic Solutions to Inhomogeneous Equations

The problem of our primary concern in this chapter is to find spectral conditions for the existence of almost periodic solutions of periodic equations. Although the theory for periodic equations can be carried out similarly to that for autonomous equations, there is always a difference between them. This is because in general there is no Floquet representation for the monodromy operators in the infinite dimensional case. Section 1 will deal with evolution semigroups acting on invariant function spaces of $AP(\mathbb{X})$. Since, originally, this technique is intended for nonautonomous equations we will treat equations with as much nonautonomousness as possible, namely, periodic equations. The spectral conditions are found in terms of spectral properties of the monodromy operators. Meanwhile, for the case of autonomous equations these conditions will be stated in terms of spectral properties of the operator coefficients. This can be done in the framework of evolution semigroups and sums of commuting operators in Section 2. Section 3 will be devoted to the critical case in which a fundamental technique of decomposition is presented. In Section 4 we will present another, but traditional, approach to periodic solutions of abstract functional differential equations. The remainder of the chapter will be devoted to several extensions of these methods to discrete systems and nonlinear equations.

3.1 Evolution Semigroups and Almost Periodic Solutions of Periodic Equations

3.1.1 *An Example*

We begin this section with an example which is a trivial case where the system has an exponential dichotomy. Let us consider the equation

$$\frac{dx}{dt} = Ax + f(t), \ t \in \mathbb{R}, x \in \mathbb{R}^n, \tag{3.1}$$

where A is an $n \times n$-matrix and f is a 1-periodic continuous function. We assume that the corresponding homogeneous equation of Eq.(3.1) has an exponential dichotomy. This is equivalent to the fact that $\sigma(A) \cap i\mathbb{R} = \oslash$, or $S^1 \cap \sigma(e^A) = \oslash$ (here S^1 denotes the unit circle in the complex plane). As is well known, in this case, for every almost periodic f, Eq.(3.1) has a unique bounded solution which can be represented in the form

$$x_f(t) = \int_{-\infty}^{t} e^{tA} P e^{-sA} f(s) ds - \int_{t}^{+\infty} e^{tA}(I - P)e^{-sA} f(s) ds, \tag{3.2}$$

where the projection P is determined from the exponential dichotomy of the homogeneous equation (P is the spectral projection from $\mathbb{R}^n$ onto the invariant subspace corresponding to the part of eigenvalues of A with negative real parts). We now prove that this bounded solution is almost periodic. In fact, we will use the Bochner's criterion. Let $\{t_k\}_{k\in \mathbf{N}}$ be any sequence of reals. We have to show that there is a subsequence $\{t_{k_l}\}$ such that the sequence $x_f(t_{k_l} + \cdot)$ is convergent to a bounded uniformly continuous function uniformly with respect to $t \in \mathbb{R}$. Note that, in this case, without loss of generality, we can assume that the projection P is commutative with the fundamental matrix e^{tA}. So, by assumption, $Pf(\cdot)$ is almost periodic. By Bochner's criterion, there is a subsequence of $\{t_{k_l}\}$ such that $Pf(t_{k_l} + \cdot)$ is convergent uniformly. Let us consider the integral

$$\int_{-\infty}^{t+t_{k_l}} e^{(t+t_{k_l}-s)A} Pf(s) ds = \int_{-\infty}^{t} e^{(t-s)A} Pf(s + t_{k_l}) ds.$$

Now from the uniform convergence of $f(t_{k_l} + \cdot)$ and the absolute convergence of the integral in (3.2) follows the uniform convergence of the above integral. Similarly, we can prove the almost periodicity of the other integral in (3.2). Hence, x_f is almost periodic.

It is interesting to find the relationship between $sp(f)$ and $sp(x_f)$. First, we assume that $f(t) = e^{i\lambda t}, \lambda \in \mathbb{R}$, and $n = 1$ (hence A is 1×1-matrix with nonzero entry). Then we have (assuming that $A = a < 0$)

$$\int_{-\infty}^{t} e^{a(t-s)} f(s) ds = \frac{1}{i\lambda - a} e^{i\lambda t}.$$

Therefore, $sp(x_f) = \{\lambda\}$. By using the Approximation Theorem, this proof can be extended to the general case of almost periodic f. In the general case we get the following inclusion:

$$sp(x_f) \subset sp(f) = \overline{\sigma_b(f)}.$$

The aim of our theory which will be presented here is to go futher in this direction. We will give conditions as sharp as possible for the existence of such almost periodic solutions as x_f. We will show that the exponential dichotomy assumption on the homogeneous equation is too strong and redundant if the spectrum of f is a strict subset of $\mathbb{R}$.

Exercise 14. Show that under the above assumption we have $\sigma_b(x_f) \subset \sigma_b(f)$.

Hint. Use Bohr transform.

3.1.2 *Evolution Semigroups*

Let us consider the following linear evolution equations

$$\frac{dx}{dt} = A(t)x, \tag{3.3}$$

and

$$\frac{dx}{dt} = A(t)x + f(t), \tag{3.4}$$

where $x \in \mathbb{X}$, $\mathbb{X}$ is a complex Banach space, $A(t)$ is a (unbounded) linear operator acting on $\mathbb{X}$ for every fixed $t \in \mathbb{R}$ such that $A(t) = A(t+1)$ for all $t \in \mathbb{R}$, $f : \mathbb{R} \to \mathbb{X}$ is an almost periodic function. Under suitable conditions Eq.(3.3) is well-posed, i.e., one can associate with equation (3.3) an evolutionary process $(U(t,s))_{t\geq s}$ which satisfies, among other things, the conditions in the following definition.

Definition 3.1. A family of bounded linear operators $(U(t,s))_{t\geq s}, (t, s \in \mathbb{R})$ from a Banach space $\mathbb{X}$ to itself is called *1-periodic strongly continuous evolutionary process* if it is a strongly continuous evolutionary process and satisfies $U(t+1, s+1) = U(t,s)$ for all $t \geq s$.

If it does not cause any danger of confusion, for the sake of simplicity, we shall often call 1-periodic strongly continuous evolutionary process *(evolutionary) process.* Note that the assumption that the period of the process is 1 is not a restriction. It is merely to shorten the notations.

Once the well-posedness of the equations in question is assumed, instead of the equations with operator-coefficient $A(t)$, we are in fact concerned with the evolutionary processes generated by these equations. In light of this, throughout the book we will deal with the asymptotic behavior of evolutionary processes as defined in Definition 3.1. Our main tool to study the asymptotic behavior of evolutionary processes is to use the notion of *evolution semigroups* associated with given evolutionary processes, which is defined in the following:

Definition 3.2. The following formal semigroup associated with a given 1-periodic strongly continuous evolutionary process $(U(t,s))_{t\geq s}$

$$(T^h u)(t) := U(t, t-h)u(t-h), \quad \forall t \in \mathbb{R},\ h \geq 0, \tag{3.5}$$

where u is an element of some function space $\mathcal{F}$, is called *evolutionary semigroup* associated with the process $(U(t,s))_{t\geq s}$ on $\mathcal{F}$.

3.1.3 *The Finite Dimensional Case*

We assume in this section that the evolutionary process $(U(t,s))_{t\geq s}$ is generated by the ordinary differential equation

$$\frac{dx}{dt} = A(t)x, \ \ t \in \mathbb{R}, x \in \mathbf{C}^n, \tag{3.6}$$

where $A(t)$ is 1-periodic continuous matrix function. Hence, by the Existence and Uniqueness Theorem, the fundamental matrix $X(t)$ associated with (3.6) exists, i.e., the matrix satisfying the Cauchy problem

$$\begin{cases} \frac{dX}{dt} = A(t)X, \ \ t \in \mathbb{R}, \\ X(0) = I. \end{cases} \tag{3.7}$$

Now setting $U(t,s) = X(t)X^{-1}(s)$ we get the so-called *Cauchy operators*, or *evolution operators* associated with (3.6). We consider the evolution semigroup $(T^h)_{h\geq 0}$ associated with this evolutionary process $(U(t,s))_{t\geq s}$ in the function space $AP(\mathbb{X})$. First, note that, by the 1-periodicity of $A(t)$ and the Existence and Uniqueness Theorem, $U(t+1, s+1) = U(t,s)$, $\forall t, s$. Thus, it is seen that

$$T^h : AP(\mathbb{X}) \to AP(\mathbb{X}), \ \forall h \geq 0.$$

Moreover, the strong continuity of the evolution semigroup can be proved easily by using classical rules of differentiation. Here we are interested in the infinitesimal generator $\mathcal{L}$ of this evolution semigroup. It is elementary to compute the generator by using the rules of differentation that is the following form: $\mathcal{L} = -d/dt + A(t)$, i.e.,

$$g \in D(\mathcal{L}) \subset AP(\mathbb{X}) \text{ if and only if } g \text{ is differentiable and } -g'(\cdot) + A(\cdot)g(\cdot) \in AP(\mathbb{X}).$$

This information suggests that the inhomogeneous equation

$$\frac{dx}{dt} = A(t)x + f(t), \ f \in AP(\mathbb{X}) \tag{3.8}$$

has an almost periodic solution $x(\cdot)$ if and only if $\mathcal{L}x = -f$. This is very useful because if we can find conditions so that the operator $\mathcal{L}$ is invertible, i.e., $0 \notin \sigma(\mathcal{L})$, then the inhomogeneous equation has a unique almost periodic solution. In turn, this condition can be found using the spectral inclusion of strongly continuous semigroups, i.e., $1 \notin \sigma(T^1)$. This section will discuss the question as how to use the above ideas for the general infinite dimensional case.

Exercise 15. Prove the formula for the generator of the evolution semigroup in the finite dimensional case.

3.1.4 *The Infinite Demensional Case*

In the infinite dimensional case many difficulties arise. First, the formula for the generator turns out to be more complicated. This is due to the fact that the Existence and Uniqueness Theorem for classical solutions does not applied to inhomogeneous equations. Hence, the classical rules of differentiation do not applied. Instead of the notion of classical solutions one introduces the one of mild solutions which saves several classical results in this case. Below we are mainly concerned with the following inhomogeneous equation

$$x(t) = U(t,s)x(s) + \int_s^t U(t,\xi)f(\xi)d\xi, \forall t \geq s \tag{3.9}$$

associated with a given strongly continuous 1-periodic evolutionary process $(U(t,s))_{t\geq s}$. A continuous solution $u(t)$ of Eq.(3.9) on an interval J will be called *mild solution* to Eq.(3.4) on J. If we do not mention the interval J for a mild solution, we mean that the mild solution is defined on the whole real line.

The following lemma will be the key tool to study spectral criteria for almost periodicity in this section which relates the generator of the evolution semigroup (3.5) with the operator defined by Eq.(3.9).

Lemma 3.1. *Let $(U(t,s))_{t\geq s}$ be a 1-periodic strongly continuous evolutionary process. Then its associated evolutionary semigroup $(T^h)_{h\geq 0}$ is strongly continuous in $AP(\mathbb{X})$. Moreover, the infinitesimal generator of $(T^h)_{h\geq 0}$ is the operator L defined as follows: $u \in D(L)$ and $Lu = -f$ if and only if $u, f \in AP(\mathbb{X})$ and u is the solution to Eq.(3.9).*

Proof. Let $v \in AP(\mathbb{X})$. First we can see that T^h acts on $AP(\mathbb{X})$. To this end, we will prove the following assertion: Let $Q(t) \in L(\mathbb{X})$ be a family of bounded linear operators which is periodic in t and strongly continuous, i.e., $Q(t)x$ is continuous in t for every given $x \in \mathbb{X}$. Then if $f(\cdot) \in AP(\mathbb{X})$, $Q(\cdot)f(\cdot) \in AP(\mathbb{X})$. The fact that $\sup_t \|Q(t\| < \infty$ follows from the Uniform Boundedness Principle. By the Approximation Theorem of almost periodic functions we can choose sequences of trigonometric polynomials $f_n(t)$ which converges uniformly to $f(t)$ on the real line. For every $n \in \mathbf{N}$, it is obvious that $Q(\cdot)f_n(\cdot) \in AP(\mathbb{X})$. Hence

$$\sup_t \|Q(t)f_n(t) - Q(t)f(t)\| \leq \sup_t \|Q(t)\| \sup_t \|f_n(t) - f(t)\|$$

implies the assertion.

We continue our proof of Lemma 3.1. By definition we have to prove that

$$\lim_{h\to 0^+} \sup_t \|U(t,t-h)v(t-h) - v(t)\| = 0. \tag{3.10}$$

Since $v \in AP(\mathbb{X})$ the range of $v(\cdot)$ which we denote by K (consisting of $x \in \mathbb{X}$ such that $x = v(t)$ for some real t) is a relatively compact subset of $\mathbb{X}$. Hence the map $(t,s,x) \mapsto U(t,s)x$ is uniformly continuous in the set $\{1 \geq t \geq s \geq -1, x \in K\}$. Now let ε be any positive real. In view of the uniform continuity of the map $(t,s,x) \mapsto U(t,s)x$ in the above-mentioned set, there is a positive real $\delta = \delta(\varepsilon)$ such that

$$\|U(t-[t], t-[t]-h)x - x\| < \varepsilon \tag{3.11}$$

for all $0 < h < \delta < 1$ and $x \in K$, where $[t]$ denotes the integer n such that $n \leq t < n+1$. Since $(U(t,s))_{t\geq s}$ is 1-periodic from (3.11) this yields

$$\lim_{h\to 0^+} \sup_t \|U(t,t-h)v(t-h) - v(t-h)\| = 0. \tag{3.12}$$

Now we have

$$\begin{aligned}
&\lim_{h\to 0^+} \sup_t \|U(t,t-h)v(t-h) - v(t)\| \\
&\qquad \leq \lim_{h\to 0^+} \sup_t \|U(t,t-h)v(t-h) - v(t-h)\| \\
&\qquad + \lim_{h\to 0^+} \sup_t \|v(t-h) - v(t)\|.
\end{aligned} \tag{3.13}$$

Since v is uniformly continuous this estimate and (3.13) imply (3.10), i.e., the evolutionary semigroup $(T^h)_{h\geq 0}$ is strongly continuous in $AP(\mathbb{X})$.

For the proof of the remainder we can do as in Section 2 of the previous chapter.

To present another proof which works also for semilinear equations, let us consider the affine semigroup $(T_f^h)_{h\geq 0}$ associated with the inhomogeneous equation (3.9) for $f \in AP(\mathbb{X})$, defined as follows:

$$T_f^h v = T^h v + \int_0^h T^{h-\xi} f d\xi = T^h v + \int_0^h T^\xi f d\xi, \tag{3.14}$$

where $v \in AP(\mathbb{X}), h \geq 0$. It is easily checked that for all $v \in AP(\mathbb{X})$ we have

$$[T_f^h v](t) = U_f(t, t-h)v(t-h), t \in \mathbb{R}, h \geq 0,$$

where $U_f(t, s)$ is the evolutionary operator defined by the integral equation (3.9). In other words, the assertion that $g, f \in AP(\mathbb{X})$ and g is a solution of (3.9), is equivalent to $T_f^h g = g$ $(\forall h \geq 0)$. From (3.14) this is equivalent to

$$g = T_f^h g = T^h g + \int_0^h T^\xi f d\xi, \quad \forall h \geq 0 \ ,$$

$$T^h g - g = \int_0^h T^\xi A g d\xi = -\int_0^h T^\xi f d\xi \ \ \forall h \geq 0. \tag{3.15}$$

From the general theory of linear operator semigroups (see Theorem 1.4) this is equivalent to the assertion $Ag = -f$. □

Remark 3.1. It may be noted that in the proof of Lemma 3.1 the precompactness of u and f are essiential. Hence, in the same way, we can show the strong continuity of the evolution semigroup $(T^h)_{h\geq 0}$ in $C_0(\mathbb{R}, \mathbb{X})$. Finally, combining this remark and Lemma 3.1 we get immediately the following corollary.

Corollary 3.1. *Let $(U(t,s))_{t\geq s}$ be a 1-periodic strongly continuous process. Then its associated evolutionary semigroup $(T^h)_{h\geq 0}$ is a C_0-semigroup in*

$$AAP(\mathbb{X}) := AP(\mathbb{X}) \oplus C_0(\mathbb{R}, \mathbb{X}).$$

One of the interesting applications of Corollary 3.1 is the following.

Corollary 3.2. *Let $(U(t,s)_{t\geq s}$ be a 1-periodic strongly continuous evolutionary process. Moreover, let $u, f \in AAP(\mathbb{X})$ such that u is a solution of Eq.(3.9). Then the almost periodic component u_{ap} of u satisfies Eq.(3.9) with $f := f_{ap}$, where f_{ap} is the corresponding almost periodic component of f.*

Proof. The evolution semigroup $(T^h)_{h\geq 0}$ leaves the subspaces $AP(\mathbb{X})$ and $C_0(\mathbb{R},\mathbb{X})$ invariant. Let us denote by P_{ap}, P_0 the projections on these function spaces, respectively. Then since u is a solution to Eq.(3.9), by Lemma 3.1,

$$\lim_{h\to 0^+}\frac{T^h u - u}{h} = -f.$$

Hence,

$$P_{ap}\lim_{h\to 0^+}\frac{T^h u - u}{h} = \lim_{h\to 0^+}\frac{T^h P_{ap}u - P_{ap}u}{h} = -P_{ap}f.$$

This, by Lemma 3.1, shows that $P_{ap}u := u_{ap}$ is a solution of Eq.(3.9) with $f := P_{ap}f := f_{ap}$. □

3.1.5 *Almost Periodic Solutions and Applications*

3.1.5.1 *Invariant functions spaces of evolution semigroups*

Below we shall consider the evolutionary semigroup $(T^h)_{h\geq 0}$ in some special invariant subspaces $\mathbf{M}$ of $AP(\mathbb{X})$.

Definition 3.3. The subspace $\mathbf{M}$ of $AP(\mathbb{X})$ is said to satisfy *condition H* if the following conditions are satisfied:

(1) $\mathbf{M}$ is a closed subspace of $AP(\mathbb{X})$,
(2) There exists $\lambda \in \mathbb{R}$ such that $\mathbf{M}$ contains all functions of the form $e^{i\lambda\cdot}x,\ x\in\mathbb{X}$,
(3) If $C(t)$ is a strongly continuous 1-periodic operator valued function and $f\in\mathbf{M}$, then $C(\cdot)f(\cdot)\in\mathbf{M}$,
(4) $\mathbf{M}$ is invariant under the group of translations.

In the sequel we will be mainly concerned with the following concrete examples of subspaces of $AP(\mathbb{X})$ which satisfy condition H:

Example 3.1. Let us denote by $\mathcal{P}(1)$ the subspace of $AP(\mathbb{X})$ consisting of all 1-periodic functions. It is clear that $\mathcal{P}(1)$ satisfies condition H.

Example 3.2. Let $(U(t,s))_{t\geq s}$ be a strongly continuous 1-periodic evolutionary process. Hereafter, for every given $f\in AP(\mathbb{X})$, we shall denote by $\mathcal{M}(f)$ the subspace of $AP(\mathbb{X})$ consisting of all almost periodic functions u such that $sp(u)\subset\overline{\{\lambda+2\pi n, n\in\mathbf{Z}, \lambda\in sp(f)\}}$. Then $\mathcal{M}(f)$ satisfies condition H.

In fact, obviously, it is a closed subspace of $AP(\mathbb{X})$, and moreover it satisfies conditions ii), iv) of the definition. We now check that condition iii) is also satisfied by proving the following lemma:

Lemma 3.2. *Let $Q(t)$ be a 1-periodic operator valued function such that the map $(t,x) \mapsto Q(t)x$ is continuous. Then for every $u(\cdot) \in AP(\mathbb{X})$, the following spectral estimate holds true:*

$$sp(Q(\cdot)u(\cdot)) \subset \Lambda, \tag{3.16}$$

where $\Lambda := \overline{\{\lambda + 2k\pi, \lambda \in sp(u), k \in \mathbf{Z}\}}$.

Proof. Using the Approximation Theorem of almost periodic functions we can choose a sequence of trignometric polynomials

$$u^{(m)}(t) = \sum_{k=1}^{N(m)} e^{i\lambda_{k,m}t} a_{k,m}, \quad a_{k,m} \in \mathbb{X}$$

such that $\lambda_{k,m} \in \sigma_b(u)$ (:= Bohr spectrum of u), $\lim_{m\to\infty} u^{(m)}(t) = u(t)$ uniformly in $t \in \mathbb{R}$. The lemma is proved if we have shown that

$$sp(Q(\cdot)u^{(m)}(\cdot)) \subset \Lambda. \tag{3.17}$$

In turn, to this end, it suffices to show that

$$sp(Q(\cdot)e^{i\lambda_{k,m}\cdot} a_{k,m}) \subset \Lambda. \tag{3.18}$$

In fact, since $Q(\cdot)a_{k,m}$ is 1-periodic in t, there is a sequence of trignometric polynomials

$$P_n(t) = \sum_{k=-N(n)}^{N(n)} e^{i2\pi kt} p_{k,n}, p_{k,n} \in \mathbb{X}$$

converging to $Q(\cdot)a_{k,m}$ uniformly as n tends to ∞. Obviously,

$$sp(e^{i\lambda_{k,m}\cdot} P_n(\cdot)) \subset \Lambda. \tag{3.19}$$

Hence,

$$sp(e^{i\lambda_{k,m}\cdot} Q(\cdot)a_{k,m}) \subset \Lambda.$$ □

An important class of invariant subspaces is that of subspaces satisfying condition H.

Proposition 3.1. *Every subspace of $AP(\mathbb{X})$ satisfying condition H is invariant under the evolution semigroup $(T^h)_{h\geq 0}$ associated with a given 1-periodic strongly continuous evolutionary process on $\mathbb{X}$.*

Proof. The proof is an easy exercise which is left to the reader. □

The following corollary will be the key tool to study the unique solvability of the inhomogeneous equation (3.9) in various subspaces $\mathbf{M}$ of $AP(\mathbb{X})$ satisfying condition H.

Corollary 3.3. *Let* $\mathbf{M}$ *satisfy condition H. Then, if* $1 \in \rho(T^1|_{\mathbf{M}})$, *the inhomogeneous equation (3.9) has a unique solution in* $\mathbf{M}$ *for every* $f \in \mathbf{M}$.

Proof. Under the assumption, the evolutionary semigroup $(T^h)_{h\geq 0}$ leaves $\mathbf{M}$ invariant. The generator A of $(T^h|_{\mathbf{M}})_{h\geq 0}$ can be defined as the part of L in $\mathbf{M}$. Thus, the corollary is an immediate consequence of Lemma 3.1 and the spectral inclusion $e^{\sigma(A)} \subset \sigma(T^1|_{\mathbf{M}})$. □

3.1.5.2 *Monodromy operators*

In view of Corollary 3.3 the problem of finding conditions for the existence of a unique almost periodic mild solution to Eq.(3.4) is now reduced to that of finding conditions for $1 \notin \sigma(T^1|_{\mathbf{M}})$. To this end, we now analyse the spectrum of $T^1|_{\mathbf{M}}$. By definition $T^1 v(t) := U(t,t-1)v(t-1)$, so it is the composition of a translation and a multiplication operator. In the theory of ordinary differential equations the operator $U(t,t-1)$ is well studied and called *monodromy operator* (more precisely, the operator $U(1,0)$).

Exercise 16. In the finite dimensional case, i.e., the process $(U(t,s))_{t\geq s}$ is the Cauchy operators of an ordinary differential equation, using the 1-periodicity and the Existence and Uniqueness Theorem show that

$$U(t,t-1) = U(t-1,0)U(1,0)U^{-1}(t-1,0) \ \forall t \in \mathbb{R}. \tag{3.20}$$

Proof. Let $U(t,0)$ be the solution of the Cauchy problem

$$\begin{cases} \frac{dU(t)}{dt} = A(t)U(t), \\ U(0) = I. \end{cases} \tag{3.21}$$

Also, the operator $U_1(t) = U(t+1,0)U^{-1}(1,0)$ is another solution to this equation due to its 1-periodicity. Using the Existence and Uniqueness Theorem for ODE we see that $U(t+1,0)U^{-1}(1,0) = U(t,0)$. Now (3.20) follows from this. □

Hence, the spectrum of $U(t,t-1)$ is the same as that of $U(1,0)$. In the infinite dimensional case, in general the process is not invertible, so this property does not hold. However, the spectral properties of $U(1,0)$ and $U(t,t-1)$ are almost the same.

First we collect some results which we shall need in the book. Recall that for a given 1-periodic evolutionary process $(U(t,s))_{t\geq s}$ the following operator

$$P(t) := U(t, t-1), t \in \mathbb{R} \tag{3.22}$$

is called *monodromy operator* (or sometime, *period map, Poincaré map*). Thus we have a family of monodromy operators. Throughout the paper we will denote $P := P(0)$. The nonzero eigenvalues of $P(t)$ are called *characteristic multipliers*. An important property of monodromy operators is stated in the following lemma.

Lemma 3.3. *Under the notation as above the following assertions hold:*

(1) $P(t+1) = P(t)$ *for all* t*; characteristic multipliers are independent of time, i.e. the nonzero eigenvalues of* $P(t)$ *coincide with those of* P*,*
(2) $\sigma(P(t))\backslash\{0\} = \sigma(P)\backslash\{0\}$*, i.e., it is independent of* t*,*
(3) If $\lambda \in \rho(P)$*, then the resolvent* $R(\lambda, P(t))$ *is strongly continuous.*

Proof. The periodicity of $P(t)$ is obvious. In view of this property we will consider only the case $0 \leq t \leq 1$. Suppose that $\mu \neq 0, Px = \mu x \neq 0$, and let $y = U(t,0)x$, so $U(1,t)y = \mu y \neq 0$, $y \neq 0$ and $P(t)y = \mu y$. By the periodicity this shows the first assertion.

Let $\lambda \neq 0$ belong to $\rho(P)$. We consider the equation

$$\lambda x - P(t)x = y, \tag{3.23}$$

where $y \in \mathbb{X}$ is given. If x is a solution to Eq.(3.23), then $\lambda x = y + w$, where $w = U(t,0)(\lambda - P)^{-1}U(1,t)y$. Conversely, defining x by this equation, it follows that $(\lambda - P(t))x = y$ so $\rho(P(t)) \supset \rho(P)\backslash\{0\}$. The second assertion follows by the periodicity. Finally, the above formula involving x proves the third assertion. □

By this lemma, once we are interested in the spectrum of $P(t)$ rather than the operators $P(t)$ for different t, we see that they are almost the same. So, by monodromy operator we may understand the operator $P := P(0)$ for convenience if this does not cause any confusion.

Let $\mathbf{M}$ be a subspace of $AP(\mathbb{X})$ invariant under the evolution semigroup $(T^h)_{h\geq 0}$ associated with the given 1-periodic evolutionary process $(U(t,s))_{t\geq s}$ in $AP(\mathbb{X})$. Below we will use the following notation

$$\hat{P}_{\mathbf{M}}v(t) := P(t)v(t), \forall t \in \mathbb{R}, v \in \mathbf{M}.$$

If $\mathbf{M} = AP(\mathbb{X})$ we will denote $\hat{P}_{\mathbf{M}} = \hat{P}$.

In the sequel we need the following lemma:

Lemma 3.4. *Let $(U(t,s))_{t\geq s}$ be a 1-periodic strongly continuous evolutionary process and* $\mathbf{M}$ *be an invariant subspace of the evolution semigroup $(T^h)_{h\geq 0}$ associated with it in $AP(\mathbb{X})$. Then for all invariant subspaces* $\mathbf{M}$ *satisfying condition H,*

$$\sigma(\hat{P}_{\mathbf{M}})\backslash\{0\} = \sigma(P)\backslash\{0\}.$$

Proof. For $u, v \in \mathbf{M}$, consider the equation $(\lambda - \hat{P}_{\mathbf{M}})u = v$. It is equivalent to the equation $(\lambda - P(t))u(t) = v(t), t \in \mathbb{R}$. If $\lambda \in \rho(\hat{P}_{\mathbf{M}})\backslash\{0\}$, for every v the first equation has a unique solution u, and $\|u\| \leq \|R(\lambda, \hat{P}_{\mathbf{M}})\|\|v\|$. Take a function $v \in \mathbf{M}$ of the form $v(t) = ye^{i\mu t}$, for some $\mu \in \mathbb{R}$; the existence of such a μ is guaranteed by the axioms of condition H. Then the solution u satisfies $\|u\| \leq \|R(\lambda, \hat{P}_{\mathbf{M}})\|\|y\|$. Hence, for every $y \in \mathbb{X}$ the solution of the equation $(\lambda - P(0))u(0) = y$ has a unique solution $u(0)$ such that

$$\|u(0)\| \leq \sup_t \|u(t)\| \leq \|R(\lambda, \hat{P}_{\mathbf{M}})\| \sup_t \|v(t)\| \leq \|R(\lambda, \hat{P}_{\mathbf{M}})\|\|y\|.$$

This implies that $\lambda \in \rho(P)\backslash\{0\}$ and $\|R(\lambda, P(t))\| \leq \|R(\lambda, \hat{P}_{\mathbf{M}})\|$.

Conversely, suppose that $\lambda \in \rho(P)\backslash\{0\}$. By Lemma 3.3 for every v the second equation has a unique solution $u(t) = R(\lambda, P(t))v(t)$ and the map taking t into $R(\lambda, P(t))$ is strongly continuous. By definition of condition H, the function taking t into $(\lambda - P(t))^{-1}v(t)$ belongs to $\mathbf{M}$. Since $R(\lambda, P(t))$ is a strongly continuous, 1-periodic function, by the uniform boundedness principle it holds that $r := \sup\{\|R(\lambda, P(t))\| : t \in R\} < \infty$. This means that $\|u(t)\| \leq r\|v(t)\| \leq r \sup_t \|v(t)\|$, or $\|u\| \leq r\|v\|$. Hence $\lambda \in \rho(\hat{P}_{\mathbf{M}})$, and $\|R(\lambda, \hat{P}_{\mathbf{M}})\| \leq r$. □

Remark 3.2. By similar argument as above we can show that if $\mathcal{P}$ is the operator of multiplication by $P(t)$ in $C_0(\mathbb{R}, \mathbb{X})$ then $\sigma(\mathcal{P})\backslash\{0\} = \sigma(P)\backslash\{0\}$. As a consequence of this we have

Corollary 3.4. *Let $(U(t,s))_{t\geq s}$ be a strongly continuous 1-periodic evolutionary process with monodromy operator P and let $\sigma(P)$ be disjoint from the unit circle. Then, the process $(U(t,s))_{t\geq s}$ has an exponential dichotomy.*

Proof. By the above remark we have that $\sigma(\mathcal{P})$ is disjoint from the unit circle as well. Let us denote the translation group in $C_0(\mathbb{R}, \mathbb{X})$ by $(S(t))_{t\in\mathbb{R}}$. Then,

$$T^1 v = \mathcal{P}S(-1)v, \quad h \geq 0, v \in C_0(\mathbb{R}, \mathbb{X}).$$

Note that $S(-1)$ and $\mathcal{P}$ are bounded linear operators that commute with each other. Therefore,

$$\sigma(T^1) = \sigma(\mathcal{P}S(-1)) \subset \sigma(\mathcal{P}) \cdot \sigma(S(-1)).$$

From the Spectral Radius Theorem it is easy to see that $\sigma(S(-1))$ is contained in (actually coincides with) the unit circle. Hence, $\sigma(T^1)$ is disjoint from the unit circle. By Theorem 2.2, the process $(U(t,s))_{t\geq s}$ has an exponential dichotomy. □

3.1.5.3 *Unique solvability of the inhomogeneous equations in* $\mathcal{P}(1)$

We now illustrate Corollary 3.2 in some concrete situations. First we will consider the unique solvability of Eq.(3.9) in $\mathcal{P}(1)$.

Proposition 3.2. *Let* $(U(t,s))_{t\geq s}$ *be 1-periodic strongly continuous. Then the following assertions are equivalent:*

(i) $1 \in \rho(P)$,

(ii) Eq.(3.9) is uniquely solvable in $\mathcal{P}(1)$ *for a given* $f \in \mathcal{P}(1)$.

Proof. Suppose that i) holds true. Then we show that ii) holds by applying Corollary 3.2. To this end, we show that $\sigma(T^1|_{\mathcal{P}(1)})\backslash\{0\} \subset \sigma(P)\backslash\{0\}$. To see this, we note that

$$T^1|_{\mathcal{P}(1)} = \hat{P}_{\mathcal{P}(1)}.$$

In view of Lemma 3.4 $1 \in \rho(T^1|_{\mathcal{P}(1)})$. By Example 3.1 and Corollary 3.2 ii) holds also true.

Conversely, we suppose that Eq.(3.9) is uniquely solvable in $\mathcal{P}(1)$. We now show that $1 \in \rho(P)$. For every $x \in \mathbb{X}$ put $f(t) = U(t,0)g(t)x$ for $t \in [0,1]$, where $g(t)$ is any continuous function of t such that $g(0) = g(1) = 0$, and

$$\int_0^1 g(t)dt = 1.$$

Thus $f(t)$ can be continued to a 1-periodic function on the real line which we denote also by $f(t)$ for short. Put $Sx = [L^{-1}(-f)](0)$. Obviously, S is a bounded operator. We have

$$[L^{-1}(-f)](1) = U(1,0)[L^{-1}(-f)](0) + \int_0^1 U(1,\xi)U(\xi,0)g(\xi)xd\xi$$

$$Sx = PSx + Px.$$

Thus

$$(I - P)(Sx + x) = Px + x - Px = x.$$

So, $I - P$ is surjective. From the uniqueness of solvability of (3.9) we get easily the injectiveness of $I - P$. In other words, $1 \in \rho(P)$. □

3.1.5.4 *Unique solvability in $AP(\mathbb{X})$ and exponential dichotomy*

This subsection will be devoted to the unique solvability of Eq.(3.9) in $AP(\mathbb{X})$ and its applications to the study of exponential dichotomy. Let us begin with the following lemma which is a consequence of Proposition 3.2.

Lemma 3.5. *Let $(U(t,s))_{t\geq s}$ be 1-periodic strongly continuous. Then the following assertions are equivalent:*

(i) $S^1 \cap \sigma(P) = \oslash$.
(ii) For every given $\mu \in \mathbb{R}, f \in \mathcal{P}(1)$ the following equation has a unique solution in $AP(\mathbb{X})$

$$x(t) = U(t,s)x(s) + \int_s^t U(t,\xi)e^{i\mu\xi}f(\xi)d\xi, \forall t \geq s. \tag{3.24}$$

Proof. Suppose that i) holds, i.e $S^1 \cap \sigma(P) = \oslash$. Then, since

$$T^1 = S(-1) \cdot \hat{P} = \hat{P} \cdot S(-1)$$

in view of the commutativeness of two operators $\hat{P}$ and $S(-1)$ we have

$$\sigma(T^1) \subset \sigma(S(-1)).\sigma(\hat{P}).$$

It may be noted that $\sigma(S(-1)) = S^1$. Thus

$$\sigma(T^1) \subset \{e^{i\mu}\lambda, \mu \in \mathbb{R}, \lambda \in \sigma(\hat{P})\}.$$

Hence, in view of Lemma 3.4

$$\sigma(T^1) \cap S^1 = \oslash.$$

Let us consider the process $(V(t,s))_{t\geq s}$ defined by

$$V(t,s)x := e^{-i\mu(t-s)}U(t,s)x$$

for all $t \geq s, x \in \mathbb{X}$. Let $Q(t)$ denote its monodromy operator, i.e. $Q(t) = e^{-i\mu}V(t,t-1)$ and $(T^h_\mu)_{h\geq 0}$ denote the evolution semigroup associated with the evolutionary process $(V(t,s))_{t\geq s}$. Then by the same argument as above we can show that since $\sigma(T^h_\mu) = e^{-i\mu}\sigma(T^h)$,

$$\sigma(T^h_\mu) \cap S^1 = \oslash.$$

By Lemma 3.1 and Corollary 3.2, the following equation

$$y(t) = V(t,s)y(s) + \int_s^t V(t,\xi)f(\xi)d\xi, \ \forall t \geq s$$

has a unique almost periodic solution $y(\cdot)$. Let $x(t) := e^{i\mu t}y(t)$. Then

$$x(t) = e^{i\mu t}y(t) = U(t,s)e^{i\mu s}y(s) + \int_s^t U(t,\xi)e^{i\mu\xi}f(\xi)d\xi$$
$$= U(t,s)x(s) + \int_s^t U(t,\xi)e^{i\mu\xi}f(\xi)d\xi \ \forall t \geq s.$$

Thus $x(\cdot)$ is an almost periodic solution of Eq.(3.24). The uniqueness of $x(\cdot)$ follows from that of the solution $y(\cdot)$.

We now prove the converse. Let $y(t)$ be the unique almost periodic solution to the equation

$$y(t) = U(t,s)y(s) + \int_s^t U(t,\xi)e^{i\mu\xi}f(\xi)d\xi, \ \forall t \geq s. \tag{3.25}$$

Then $x(t) := e^{-i\mu t}y(t)$ must be the unique solution to the following equation

$$x(t) = e^{-i\mu(t-s)}U(t,s)x(s) + \int_s^t e^{-i\mu(t-\xi)}U(t,\xi)f(\xi)d\xi), \ \forall t \geq s. \tag{3.26}$$

And vice versa. We show that $x(t)$ should be periodic. In fact, it is easily seen that $x(1+\cdot)$ is also an almost periodic solution to Eq.(3.25). From the uniqueness of $y(\cdot)$ (and then that of $x(\cdot)$) we have $x(t+1) = x(t)$, $\forall t$. By Proposition 3.2 this yields that $1 \in \rho(Q(0))$, or in other words, $e^{i\mu} \in \rho(P)$. From the arbitrary nature of μ, $S^1 \cap \sigma(P) = \oslash$. □

Theorem 3.1. *Let $(U(t,s))_{t\geq s}$ be given 1-periodic strongly continuous evolutionary process. Then the following assertions are equivalent:*

(i) The process $(U(t,s))_{t\geq s}$ has an exponential dichotomy;
(ii) For every given bounded and continuous f the inhomogeneous equation (3.9) has a unique bounded solution;
(iii) The spectrum of the monodromy operator P does not intersect the unit circle;
(iv) For every given $f \in AP(\mathbb{X})$ the inhomogeneous equation (3.9) is uniquely solvable in the function space $AP(\mathbb{X})$.

Proof. The equivalence of i) and ii) has been established in Theorem 2.1 and the remarks that follow. Now we show the equivalence between i), ii) and iii). The fact that iii) implies i) is the content of Corollary 3.4. Now we prove that i) implies iii). Let the process have an exponential dichotomy. We now show that the spectrum of the monodromy operator P does not intersect the unit circle. We will follow the argument of Lemma 3.5. Since

i) is equivalent to ii), for every fixed real number μ and $f \in \mathcal{P}(1)$, there exists a unique bounded solution y to (3.25). By the argument of Lemma 3.5, the function $x(t) := e^{-i\mu t}y(t)$ must be the unique bounded solution to (3.26) and that should be 1-periodic. Therefore, actually the solution y is almost periodic. By Lemma 3.5, $S^1 \cap \sigma(P) = \emptyset$. So i) is equivalent to iii).

It remains to show the equivalence between iii) and iv). We first show the implication *iii)* $\Rightarrow$ *iv)*. By Lemma 3.4 the operator $\hat{P}$ of multiplication by $P(t)$ on $AP(\mathbb{X})$ has the property that $\sigma(\hat{P}) \cap S^1 = \emptyset$. Therefore, since $\hat{P}$ commutes with translation operator $S(t)$ the evolution semigroup T^h on $AP(\mathbb{X})$ has the property that

$$\sigma(T^1) = \sigma(\hat{P}S(-1)) \subset \sigma(\hat{P}) \cdot \sigma(S(-1)).$$

Since $\sigma(S(-1)) = S^1$ we have that $\sigma(T^1) \cap S^1 = \emptyset$. By the Spectral Inclusion Theorem for C_0-semigroup and Lemma 3.1 we have iv). Conversely, let iv) hold. Then by Lemma 3.5 we have iii). □

3.1.5.5 *Unique solvability of the inhomogeneous equations in $\mathcal{M}(f)$*

Now let us return to the more general case where the spectrum of the monodromy operator may intersect the unit circle.

Theorem 3.2. *Let $(U(t,s))_{t\geq s}$ be a 1-periodic strongly continuous evolutionary process. Moreover, let $f \in AP(\mathbb{X})$ such that $\sigma(P) \cap \overline{\{e^{i\lambda}, \lambda \in sp(f)\}} = \oslash$. Then the inhomogeneous equation (3.9) has an almost periodic solution which is unique in $\mathcal{M}(f)$.*

Proof. From Example 3.2 it follows that the function space $\mathcal{M}(f)$ satisfies condition H. Since $(S(t))_{t\in\mathbb{R}}$ is an isometric C_0-group, by the weak spectral mapping theorem for isometric groups (see e.g. [Nagel (73)]) we have

$$\sigma(S(1)|_{\mathcal{M}(f)}) = \overline{e^{\sigma(\mathcal{D}|_{\mathcal{M}(f)})}},$$

where $\mathcal{D}|_{\mathcal{M}(f)}$ is the generator of $(S(t)|_{\mathcal{M}(f)})_{t\geq 0}$. From the general spectral theory of bounded functions we have

$$\sigma(\mathcal{D}|_{\mathcal{M}(f)}) = i\Lambda,$$

where $\Lambda = \overline{\{\lambda + 2\pi k, \lambda \in sp(f), k \in \mathbf{Z}\}}$. Hence, since

$$e^{\sigma(\mathcal{D}|_{\mathcal{M}(f)})} = e^{i\Lambda} \subset \overline{e^{isp(f)}} \subset \overline{e^{i\Lambda}},$$

we have

$$\sigma(S(1)|_{\mathcal{M}(f)}) = \overline{e^{\sigma(\mathcal{D}|_{\mathcal{M}(f)})}} = \overline{e^{isp(f)}} \ .$$

Thus, the condition

$$\sigma(P) \cap \overline{e^{isp(f)}} = \oslash$$

is equivalent to the following

$$1 \notin \sigma(P).\sigma(S(-1)|_{\mathcal{M}(f)}).$$

In view of the inclusion

$$\begin{aligned}\sigma(T^1|_{\mathcal{M}(f)})\backslash\{0\} &\subset \sigma(\hat{P}_{\mathcal{M}(f)}).\sigma(S(-1)|_{\mathcal{M}(f)})\backslash\{0\} \\ &\subset \sigma(P).\sigma(S(-1)|_{\mathcal{M}(f)})\backslash\{0\}\end{aligned}$$

which follows from the commutativeness of the operator $\hat{P}_{\mathcal{M}(f)}$ with the operator $S(-1)|_{\mathcal{M}(f)}$, the above inclusion implies that

$$1 \notin \sigma(T^1|_{\mathcal{M}(f)}).$$

Now the assertion of the theorem follows from Corollary 3.2. □

The inverse of the above theorem is the following:

Theorem 3.3. *Let $(U(t,s))_{t\geq s}$ be a 1-periodic strongly continuous evolutionary process. Moreover, let the inhomogeneous equation (3.9) have a unique almost periodic solution in $\mathcal{M}(f)$. Then $\sigma(P) \cap \{e^{i\lambda}, \lambda \in sp(f)\} = \oslash$.*

Proof. Let $\mu \in sp(f)$. Consider the function $f_\lambda(t) := e^{i\mu t}f(t)$, where $f(t) \in \mathcal{P}(1)$. Then there is a unique solution $x_\mu(\cdot) \in AP(\mathbb{X})$ of the equation (3.24), or equivalently, a unique solution $y(t) = e^{-i\mu t}x(t)$ to the equation (3.26). Since, the equation (3.26) is periodic in t, if there is a solution $y(\cdot)$, then $y(1+\cdot)$ is also a solution. From the uniqueness it follows that $y(1+\cdot) = y(\cdot)$. This means that $y(\cdot)$ is 1-periodic. Hence, for every $f \in \mathcal{P}(1)$ Eq.(3.26) has a unique 1-periodic solution $y(\cdot)$. By Proposition 3.2 $0 \notin \sigma(Q(0))$ which is equivalent to $e^{i\mu} \notin \sigma(P)$. Hence, $\{e^{i\mu}, \mu \in sp(f)\} \cap \sigma(P) = \oslash$. □

3.1.5.6 *Unique solvability of nonlinearly perturbed equations*

Let us consider the semilinear equation

$$x(t) = U(t,s)x(s) + \int_s^t U(t,\xi)g(\xi, x(\xi))d\xi. \tag{3.27}$$

We shall be interested in the unique solvability of (3.27) for a larger class of the forcing term g. We shall show that the generator of evolutionary semigroup is still useful in studying the perturbation theory in the critical case in which the spectrum of the monodromy operator P may intersect the unit circle. We suppose that $g(t,x)$ is Lipschitz continuous with coefficient k and the Nemystky operator F defined by $(Fv)(t) = g(t, v(t)), \forall t \in \mathbb{R}$ acts in $\mathbf{M}$. Below we can assume that $\mathbf{M}$ *is any closed subspace of* the space of all bounded continuous functions $BC(\mathbb{R}, \mathbb{X})$. We consider the operator L in $BC(\mathbb{R}, \mathbb{X})$. If $(U(t,s))_{t \geq s}$ is strongly continuous, then L is a single-valued operator from $D(L) \subset BC(\mathbb{R}, \mathbb{X})$ to $BC(\mathbb{R}, \mathbb{X})$.

Lemma 3.6. *Let* $\mathbf{M}$ *be any closed subspace of* $BC(\mathbb{R}, \mathbb{X})$, $(U(t,s))_{t \geq s}$ *be strongly continuous and Eq.(3.9) be uniquely solvable in* $\mathbf{M}$. *Then for sufficiently small* k , *Eq.(3.27) is also uniquely solvable in this space.*

Proof. First, we observe that under the assumptions of the lemma we can define a single-valued operator L acting in $\mathbf{M}$ as follows: $u \in D(L)$ if and only if there is a function $f \in \mathbf{M}$ such that Eq.(3.9) holds. From the strong continuity of the evolutionary process $(U(t,s))_{t \geq s}$ one can easily see that there is at most one function f such that Eq.(3.9) holds. This means L is single-valued. Moreover, one can see that L is closed. Now we consider the Banach space $[D(L)]$ with graph norm, i.e. $|v| = \|v\| + \|Lv\|$. By assumption it is seen that L is an isomorphism from $[D(L)]$ onto $\mathbf{M}$. In view of the Lipschitz Inverse Mapping Theorem for Lischitz mappings for sufficiently small k the operator $L - F$ is invertible. Hence there is a unique $u \in \mathbf{M}$ such that $Lu - Fu = 0$. From the definition of operator L we see that u is a unique solution to Eq.(3.27). □

Corollary 3.5. *Let* $\mathbf{M}$ *be any closed subspace of* $AP(\mathbb{X})$, $(U(t,s))_{t \geq s}$ *be 1-periodic strongly continuous evolutionary process and for every* $f \in \mathbf{M}$ *the inhomogeneous equation (3.9) be uniquely solvable in* $\mathbf{M}$. *Moreover let the Nemytsky operator* F *induced by the nonlinear function* g *in Eq.(3.27) act on* $\mathbf{M}$. *Then for sufficiently small* k , *the semilinear equation (3.27) is uniquely solvable in* $\mathbf{M}$.

Proof. The corollary is an immediate consequence of Lemma 3.6. □

3.1.5.7 *Example 1*

In this example we shall consider the abstract form of parabolic partial differential equations (see e.g. [Henry (46)]) and apply the results obtained above to study the existence of almost periodic solutions to these equations. It may be noted that a necessary condition for the existence of Floquet representation is that the process under consideration is invertible. It is known for the bounded case (see e.g. Chap. V, Theorem 1.2 in [Daleckii and Krein (24)]) that if the spectrum of the monodromy operator does not circle the origin (of course, it should not contain the origin), then the evolution operators admit Floquet representation. In the example below, in general, Floquet representation does not exist. For instance, if the sectorial operator A has compact resolvent, then monodromy operator is compact (see [Henry (46)] for more details). Thus, if $dim\mathbb{X} = \infty$, then monodromy operators cannot be invertible. However, the above results can apply.

Let A be sectorial operator in a Banach space $\mathbb{X}$, and the mapping taking t into $B(t) \in L(\mathbb{X}^\alpha, \mathbb{X})$ be Hölder continuous and 1-periodic. Then there is a 1-periodic evolutionary process $(U(t,s))_{t\geq s}$ associated with the equation

$$\frac{du}{dt} = (-A + B(t))u. \tag{3.28}$$

We have the following:

Claim 1 *For any $x_0 \in \mathbb{X}$ and τ there exists a unique (strong) solution $x(t) := x(t;\tau,x_0)$ of Eq.(3.28) on $[\tau, +\infty)$ such that $x(\tau) = x_0$. Moreover, if we write $x(t;\tau,x_0) := T(t,\tau)x_0, \forall t \geq \tau$, then $(T(t,\tau))_{t\geq\tau}$ is a strongly continuous 1-periodic evolutionary process. In addition, if A has compact resolvent, then the monodromy operator $P(t)$ is compact.*

Proof. This claim is an immediate consequence of Theorem 7.1.3, pp. 190-191 in [Henry (46)]. In fact, it is clear that $(T(t,\tau))_{t\geq\tau}$ is strongly continuous and 1-periodic. The last assertion is contained in Lemma 7.2.2, p. 197 in [Henry (46)]). □

Thus, in view of the above claim if $dim\mathbb{X} = \infty$, then Floquet representation does not exist for the process. This means the problem cannot reduced to the autonomous and bounded case. To apply our results, let the function f taking t into $f(t) \in \mathbb{X}$ be almost periodic and the spectrum of the monodromy operator of the process $(U(t,s))_{t\geq s}$ be separated from the set $\overline{e^{isp(f)}}$. Then the following inhomogeneous equation

$$\frac{du}{dt} = (-A + B(t))u + f(t)$$

has a unique almost periodic solution u such that

$$sp(u) \subset \overline{\{\lambda + 2\pi k, k \in \mathbf{Z}, \lambda \in sp(f)\}}.$$

We now show

Claim 2 *Let the conditions of Claim 1 be satisfied except for the compactness of the resolvent of A . Then*

$$\frac{dx}{dt} = (-A + B(t))x \tag{3.29}$$

has an exponential dichotomy if and only if the spectrum of the monodromy operator does not intersect the unit circle. Moreover, if A has compact resolvent, it has an exponential dichotomy if and only if all multipliers have modulus different from one. In particular, it is asymptotically stable if and only if all characteristic multipliers have modulus less than one.

Proof. The operator $T(t,s), t > s$ is compact if A has compact resolvent (see e.g. p. 196 in [Henry (46)]). The claim is an immediate consequence of Theorem 3.1. □

3.1.5.8 *Example 2*

We examine in this example how the condition of Theorem 3.2 cannot be dropped. In fact we consider the simplest case with $A = 0$

$$\frac{dx}{dt} = f(t), \quad x \in \mathbb{R}, \tag{3.30}$$

where f is continuous and 1-periodic. Obviously,

$$\sigma(e^A) = \{1\} = \overline{e^{i\ sp(f)}}.$$

We assume further that the integral $\int_0^t f(\xi)d\xi$ is bounded. Then every solution to Eq.(3.30) can be extended to a periodic solution defined on the whole line of the form

$$x(t) = c + \int_0^t f(\xi)d\xi, t \in \mathbb{R}.$$

Thus the uniqueness of a periodic solution to Eq.(3.30) does not hold.

Now let us consider the same Eq.(3.30) but with 1-anti-periodic f, i.e., $f(t+1) = f(t), \forall t \in \mathbb{R}$. Clearly,

$$\overline{e^{i\ sp(f)}} = \{-1\} \cap \sigma(e^A) = \emptyset.$$

Hence the conditions of Theorem 3.2 are satisfied. Recall that in this theorem we claim that the uniqueness of the almost periodic solutions is among the class of almost periodic functions g with $\overline{e^{i\ sp(g)}} \subset \overline{e^{i\ sp(f)}}$. Now let us have a look at our example. Every solution to Eq.(3.30) is a sum of the unique 1-anti-periodic solution, which existence is guaranteed by Theorem 3.2, and a solution to the corresponding homogeneous equation, i.e., in this case a constant function. Hence, Eq.(3.30) has infinitely many almost periodic solutions.

3.2 Evolution Semigroups and Sums of Commuting Operators

Let $\mathbb{X}$ be a given complex Banach space and $\mathcal{M}$ be a translation invariant subspace of the space of $\mathbb{X}$-valued bounded uniformly continuous functions on the real line (that is denoted by $BUC(\mathbb{R}, \mathbb{X})$). The problem we consider in this section is to find conditions for $\mathcal{M}$ to be admissible with respect to differential equations of the form

$$\frac{du}{dt} = Au + f(t), \tag{3.31}$$

where A is an (unbounded) linear operator with nonempty resolvent set on the Banach space $\mathbb{X}$. By a tradition, by admissibility here we mean that for every $f \in \mathcal{M}$ Eq.(3.31) has a unique solution (in a suitable sense) which belongs to $\mathcal{M}$ as well. The main condition obtained in this section is of the form

$$\sigma(A) \cap isp(f) = \emptyset.$$

We will show that the method of sums of commuting operators can be extended to larger classes of equations, including abstract functional differential equations.

3.2.1 *Invariant Function Spaces*

By $(S(t))_{t\in\mathbb{R}}$ we denote the translation group on the function space $BUC(\mathbb{R}, \mathbb{X})$, i.e., $S(t)v(s) := v(t+s), \forall t, s \in \mathbb{R}, v \in BUC(\mathbb{R}, \mathbb{X})$ with infinitesimal generator $\mathcal{D} := d/dt$ defined on $D(\mathcal{D}) := BUC^1(\mathbb{R}, \mathbb{X})$. Let $\mathcal{M}$ be a subspace of $BUC(\mathbb{R}, \mathbb{X})$, and let A be a linear operator on $\mathbb{X}$. We will denote by $\mathcal{A}_{\mathcal{M}}$ the operator $f \in \mathcal{M} \mapsto Af(\cdot)$ with $D(\mathcal{A}_{\mathcal{M}}) = \{f \in \mathcal{M} | \forall t \in \mathbb{R}, f(t) \in D(A), Af(\cdot) \in \mathcal{M}\}$. When $\mathcal{M} = BUC(\mathbb{R}, \mathbb{X})$ we will use the notation $\mathcal{A} := \mathcal{A}_{\mathcal{M}}$. Throughout the paragraph we always assume that A is a given operator on $\mathbb{X}$ with $\rho(A) \neq \oslash$, (so it is closed).

In this paragraph we will use the notion of translation-invariance of a function space, which we recall in the following definition, and additional conditions on it.

Definition 3.4. A closed and translation invariant subspace $\mathcal{M}$ of the function space
$BUC(\mathbb{R}, \mathbb{X})$, i.e., $S(\tau)\mathcal{M} \subset \mathcal{M}$ for all $\tau \in \mathbb{R}$, is said to satisfy

(1) *condition H1* if the following condition is fulfilled:
$$\forall C \in L(\mathbb{X}), \forall f \in \mathcal{M} \Rightarrow Cf \in \mathcal{M},$$
(2) *condition H2* if the following condition is fulfilled:
For every closed linear operator A, if $f \in \mathcal{M}$ such that $f(t) \in D(A)$, $\forall t$, $\mathcal{A}f \in BUC(\mathbb{R}, \mathbb{X})$, then $\mathcal{A}f \in \mathcal{M}$,
(3) *condition H3* if the following condition is fulfilled: For every bounded linear operator $B \in L(BUC(\mathbb{R}, \mathbb{X}))$ which commutes with the translation group $(S(t))_{t\in\mathbb{R}}$ one has $B\mathcal{M} \subset \mathcal{M}$.

Remark 3.3. As remarked in p.401 in [Vu and Schuler (103)], condition H3 is equivalent to the assertion that
$$\forall B \in L(\mathcal{M}, \mathbb{X}) \ \ \forall f \in \mathcal{M} \Rightarrow BS(\cdot)f \in \mathcal{M}.$$
Obviously, conditions H2, H3 are stronger than condition H1. In the sequel, we will define the autonomousness of a functional operator via condition H3.

Example 3.3. If $\Lambda(\mathbb{X}) = \{f \in BUC(\mathbb{R}, \mathbb{X}) : sp(f) \subset \Lambda\}$, where Λ is a given closed subset of $\mathbb{R}$. Then $\Lambda(\mathbb{X})$ is a translation invariant closed subspace of $BUC(\mathbb{R}, \mathbb{X})$. Moreover, it satisfies all conditions H1, H2, H3.

In connection with the translation-invariant subspaces we need the following simple spectral properties.

Lemma 3.7.

(i) Let $\mathcal{M}$ satisfy condition H1. Then
$$\sigma(\mathcal{A}_{\mathcal{M}}) \subset \sigma(\mathcal{A}) = \sigma(A)$$
and
$$\|R(\lambda, \mathcal{A}_{\mathcal{M}})\| \le \|R(\lambda, \mathcal{A})\| = \|R(\lambda, A)\|, \forall \lambda \in \rho(A);$$
(ii) Let $\mathcal{M}$ satisfy condition H3 and $\mathcal{B}$ be a bounded linear operator on $BUC(\mathbb{R}, \mathbb{X})$ which commutes with the translation group. Then $\sigma(\mathcal{B}_{\mathcal{M}}) \subset \sigma(\mathcal{B})$ and
$$\|R(\lambda, \mathcal{B}_{\mathcal{M}})\| \le \|R(\lambda, \mathcal{B})\|, \forall \lambda \in \rho(\mathcal{B}).$$

Proof. i) Let $\lambda \in \rho(A)$. We show that $\lambda \in \rho(\mathcal{A}_{\mathcal{M}})$. In fact, as $\mathcal{M}$ satisfies condition H1, $\forall f \in \mathcal{M}, R(\lambda, A)f(\cdot) := (\lambda - A)^{-1}f(\cdot) \in \mathcal{M}$. Thus the function $R(\lambda, A)f(\cdot)$ is a solution to the equation $(\lambda - \mathcal{A}_{\mathcal{M}})u = f$. Moreover, since $\lambda \in \rho(A)$ it is seen that the above equation has at most

one solution. Hence $\lambda \in \rho(\mathcal{A}_{\mathcal{M}})$. Moreover, it is seen that $\|R(\lambda, \mathcal{A}_{\mathcal{M}})\| \leq \|R(\lambda, A)\|$. Similarly, we can show that if $\lambda \in \rho(\mathcal{A})$, then $\lambda \in \rho(A)$ and $\|R(\lambda, A)\| \leq \|R(\lambda, \mathcal{A})\|$.
ii) The proof of the second assertion can be done in the same way. □

In the section, as a model of the translation - invariant subspaces, which satisfy all conditions H1, H2, H3 we can take the spectral spaces

$$\Lambda(\mathbb{X}) := \{u \in BUC(\mathbb{R}, \mathbb{X}) : \mathrm{sp(u)} \subset \Lambda\},$$

where Λ is a given closed subset of the real line.

3.2.2 *Differential Operator $d/dt - \mathcal{A}$ and Notions of Admissibility*

We start the main subsection of this section by discussing various notions of admissibility and their inter-relations via the differential operator $d/dt - \mathcal{A}$, or more precisely its closed extensions, for the following equation

$$\frac{dx}{dt} = Ax + f(t), x \in \mathbb{X}, t \in \mathbb{R}, \tag{3.32}$$

where A is a linear operator acting on $\mathbb{X}$.

We first recall that

Definition 3.5.

(1) An $\mathbb{X}$-valued function u on $\mathbb{R}$ is said to be a *solution on* $\mathbb{R}$ to Eq.(3.32) for given linear operator A and $f \in BUC(\mathbb{R}, \mathbb{X})$ (or sometime, *classical solution*) if $u \in BUC^1(\mathbb{R}, \mathbb{X}), u(t) \in D(A), \forall t$ and u satisfies Eq.(3.32) for all $t \in \mathbb{R}$.
(2) Let A be the generator of a C_0 semigroup of linear operators. An $\mathbb{X}$-valued continuous function u on $\mathbb{R}$ is said to be a *mild solution on* $\mathbb{R}$ to Eq.(3.32) for a given $f \in BUC(\mathbb{R}, \mathbb{X})$ if u satisfies

$$u(t) = e^{(t-s)A}u(s) + \int_s^t e^{(t-r)A} f(r)dr, \forall t \geq s.$$

Definition 3.6.

(1) A closed translation invariant subspace $\mathcal{M} \subset BUC(\mathbb{R}, \mathbb{X})$ is said to be *admissible for Eq.(3.32)* if for each $f \in \mathcal{M}_0 := \mathcal{M} \cap BUC^1(\mathbb{R}, \mathbb{X})$ there is a unique solution $u \in \mathcal{M}_0$ of Eq.(3.32) and if $f_n \in \mathcal{M}_0, n \in \mathbf{N}, f_n \to 0$ as $n \to \infty$ in $\mathcal{M}_0$ imply $u_n \to 0$ as $n \to \infty$.

(2) Let A be the generator of a C_0-semigroup. A translation - invariant closed subspace $\mathcal{M}$ of $BUC(\mathbb{R},\mathbb{X})$ is said to be *mildly admissible for Eq.(3.32)* if for every $f \in \mathcal{M}$ there exists a unique mild solution $x_f \in \mathcal{M}$ to Eq.(3.32).

Exercise 17. By definition it is obvious that admissibility implies that if $\mathcal{D}_{\mathcal{M}} - \mathcal{A}_{\mathcal{M}}$ is closable, one has $0 \in \rho(\overline{\mathcal{D}_{\mathcal{M}} - \mathcal{A}_{\mathcal{M}}})$.

Proof. Set $\mathcal{B} := \mathcal{D}_{\mathcal{M}} - \mathcal{A}_{\mathcal{M}}$. Obviously, $\mathcal{B}$ is injective. In fact, if $\mathcal{B}f = 0$ then from the uniqueness assumption $f = 0$. By assumption, from the continuity of the Green operator G, defined as the unique extension of $f \in \mathcal{M}_0 \mapsto u \in \mathcal{M}_0$ it follows that there exists $m > 0$ such that $\|\mathcal{B}x\| \geq \|x\|$ for all $x \in D(\mathcal{B})$. We now show that $\overline{\mathcal{B}}$ is injective. Indeed, it sufficies to show that if $\overline{\mathcal{B}}y = 0$, then $y = 0$. By definition, there exist $(y_n, x_n) \in \Gamma(\mathcal{B})$ such that $y_n \to y$, $\mathcal{B}y_n = x_n \to 0 = \mathcal{B}y$. Hence, $0 = \lim n \to \infty \|x_n\| = \|\mathcal{B}y\| \geq m\|y\|$, so $y = 0$. We show that $\overline{\mathcal{B}}$ is surjective. Let $f \in \mathcal{M}$. We show that Gf is in $D(\overline{\mathcal{B}}$ and $\overline{\mathcal{B}}Gf = f$. In fact, by definition of the extension G, there are $f_n \in \mathcal{M}_0$ such that $f_n \to f$ and $Gf_n \to Gf$. Obviously, $\overline{\mathcal{B}}Gf_n = \mathcal{B}Gf_n = f_n$. This completes the proof of the exrcise. □

We now discuss the relationship between the notions of admissibility, weak admissibility and mild admissibility if A is the generator of a C_0-semigroup. To this end, we introduce the following operator $L_{\mathcal{M}}$ which will be the key tool in our construction.

Definition 3.7. Let $\mathcal{M}$ be a translation invariant closed subspace of $BUC(\mathbb{R},\mathbb{X})$. We define the operator $L_{\mathcal{M}}$ on $\mathcal{M}$ as follows: $u \in D(L_{\mathcal{M}})$ if and only if $u \in \mathcal{M}$ and there is $f \in \mathcal{M}$ such that

$$u(t) = e^{(t-s)A}u(s) + \int_s^t e^{(t-r)A} f(r)dr, \forall t \geq s \tag{3.33}$$

and in this case $L_{\mathcal{M}}u := f$.

The following lemma will be needed in the sequel

Lemma 3.8. *Let A be the generator of a C_0-semigroup and $\mathcal{M}$ be a closed translation invariant subspace of $AAP(\mathbb{X})$ which satisfies condition H1. Then*

$$\overline{\mathcal{D}_{\mathcal{M}} - \mathcal{A}_{\mathcal{M}}} = L_{\mathcal{M}}.$$

Proof. Let us consider the semigroup $(T^h)_{h\geq 0}$

$$T^h v(t) := e^{hA}v(t-h), v \in \mathcal{M}, h \geq 0.$$

By condition H1, clearly, $(T^h)_{h\geq 0}$ leaves $\mathcal{M}$ invariant. By Corollary 3.1, since $\mathcal{M} \subset AAP(\mathbb{X})$ this semigroup is strongly continuous which has $-L_{\mathcal{M}}$ as its generator. On the other hand, since $(T^h)_{h\geq 0}$ is the composition of two commuting and strongly continuous semigroups, by p. 24 in [Nagel (73)] this generator is nothing but $\overline{-\mathcal{D}_{\mathcal{M}} + \mathcal{A}_{\mathcal{M}}}$. □

Corollary 3.6. *Let A be the generator of a C_0-semigroup and $\mathcal{M}$ be a translation invariant closed subspace of $BUC(\mathbb{R}, \mathbb{X})$. Then the notions of admissibility and mild admissibility of $\mathcal{M}$ for Eq.(3.32) are equivalent provided that $\mathcal{M}$ satisfies condition H1 and $\mathcal{M} \subset AAP(\mathbb{X})$.*

Proof. Since, by Exercise 17 the admissibility of $\mathcal{M}$ for Eq.(3.32) implies in particular that $0 \in \rho(\overline{\mathcal{D}_{\mathcal{M}} - \mathcal{A}_{\mathcal{M}}})$, and by Lemma 3.8

$$\overline{\mathcal{D}_{\mathcal{M}} - \mathcal{A}_{\mathcal{M}}} = \overline{\mathcal{D}_{\mathcal{M}} - \mathcal{A}_{\mathcal{M}}}^A = L_{\mathcal{M}}$$

the implication "*admissibility ⇒ mild admissibility*" is clear. *It remains only to show "mild admissibility ⇒ admissibility", i.e., if*

$$0 \in \rho(L_{\mathcal{M}}),$$

then $\mathcal{M}$ is admissible with respect to Eq.(3.32). In fact, by assumption, for every $f \in \mathcal{M}$ there is a unique mild solution $u := L^{-1}_{\mathcal{M}} f$ of Eq.(3.32). It can be seen that the function $u(\tau + \cdot) \in \mathcal{M}$ is a mild solution of Eq.(3.32) with the forcing term $f(\tau + \cdot)$ for every fixed $\tau \in \mathbb{R}$. Hence, by the uniqueness, $u(\tau + \cdot) = L^{-1}_{\mathcal{M}} f(\tau + \cdot)$. We can rewrite this fact as

$$S(\tau)L^{-1}_{\mathcal{M}} f = L^{-1}_{\mathcal{M}} S(\tau) f, \forall f \in \mathcal{M}, \tau \in \mathbb{R}.$$

From this and the boundedness of $L^{-1}_{\mathcal{M}}$,

$$\lim_{\tau\to 0^+} \frac{S(\tau)u - u}{\tau} = L^{-1}_{\mathcal{M}} \lim_{\tau\to 0^+} \frac{S(\tau)f - f}{\tau}.$$

Thus, the assumption that $f \in \mathcal{M}_0$ implies that the left hand side limit exists. Thus, $u = L^{-1}_{\mathcal{M}} f \in \mathcal{M}_0$. As is well known, since f is differentiable $\int_s^t e^{(t-\xi)A} f(\xi) d\xi$ is differentiable (see Theorem, p. 84 in [Goldstein (36)]). Thus, by definition of mild solutions, from the differentiability of u it follows that $e^{(t-s)A}u(s)$ is differentiable with respect to $t \geq s$. Thus, $u(s) \in D(A)$ for every $s \in \mathbb{R}$. Finally, this shows that $u(\cdot)$ is a classical solution to Eq.(3.32) on $\mathbb{R}$. Hence the admissibility of $\mathcal{M}$ for Eq.(3.32) is proved. □

3.2.3 *Admissibility for Abstract Ordinary Differential Equations*

In this subsection we shall demonstrate some advantages of using the operator $d/dt - \mathcal{A}$ as the sum of two commuting operators to study the admissibility theory for Eq.(3.32).

In the sequel we shall need the following basic property of the translation group on $\Lambda(\mathbb{X})$ which proof can be done in a standard manner.

Lemma 3.9. *Let Λ be a closed subset of the real line. Then*

$$\sigma(\mathcal{D}_{\Lambda(\mathbb{X})}) = i\Lambda.$$

Proof. First, we note that for every $\lambda \in \Lambda, i\lambda \in \sigma(\mathcal{D}_{\Lambda(\mathbb{X})})$. In fact, $\mathcal{D}e^{i\lambda \cdot}x = i\lambda e^{i\lambda \cdot}x$. Now suppose that $\lambda_0 \notin \Lambda$. Then we shall show that $i\lambda_0 \in \rho(\mathcal{D}_{\Lambda(\mathbb{X})})$. To this end, we consider the following equation

$$\frac{du}{dt} = i\lambda_0 u + g(t), g \in \Lambda(\mathbb{X}). \tag{3.34}$$

Since $isp(g) = \sigma(\mathcal{D}_{\mathcal{M}_g})$, where $\mathcal{M}_g$ is the closed subspace of $BUC(\mathbb{R},\mathbb{X})$, spanned by all translations of g (see e.g., [Arendt and Batty (5)],[Vu (101)]), we get $i\lambda_0 \notin \sigma(\mathcal{D}_{\mathcal{M}_g})$; and hence the above equation has a unique solution $h \in \mathcal{M}_g \subset \Lambda(\mathbb{X})$. If k is another solution to Eq.(3.34) in $\Lambda(\mathbb{X})$, then $h - k$ is a solution in $\Lambda(\mathbb{X})$ to the homogeneous equation associated with Eq.(3.34). Thus, a computation via Carleman transform shows that $sp(h - k) \subset \{\lambda_0\}$. On the one hand, we get $\lambda_0 \notin sp(h - k)$ because of $sp(h - k) \subset \Lambda$. Hence, $sp(h - k) = \oslash$, and then $h - k = 0$. In other words, Eq.(3.34) has a unique solution in $\Lambda(\mathbb{X})$. This shows that the above equation has a unique solution in $\Lambda(\mathbb{X})$, i.e. $i\lambda_0 \in \rho(\mathcal{D}_{\Lambda(\mathbb{X})})$. □

Theorem 3.4. *Let A be the generator of a C_0-semigroup $(T(t))_{t\geq 0}$ of linear operators on $\mathbb{X}$ and let $\Lambda(\mathbb{X})$ be mildly admissible for Eq.(3.32). Then,*

$$i\Lambda \cap \sigma(A) = \oslash.$$

Proof. Suppose that G is the operator which takes every $g \in \Lambda(\mathbb{X})$ into the unique mild solution u_f of Eq. (3.32). We show that G commutes with the translation group $(S(t))_{t\in\mathbb{R}}$. In fact, for every $\tau \in \mathbb{R}$ we see that if $f \in \Lambda(\mathbb{X})$, then

$$\begin{aligned}[Gf](\tau + t) &= T((\tau + t) - (\tau + s))[Gf](\tau + s) \\ &\quad + \int_{\tau+s}^{\tau+t} T(\tau + t - \xi)f(\xi)d\xi, \quad \forall t \geq s \\ &= T(t - s)[Gf](\tau + s) + \int_s^t T(t - \eta)f(\tau + \eta)d\eta,\end{aligned}$$

and hence, the function $u(t) = [Gf](\tau + t) = [S(\tau)Gf](t)$ is the unique solution to Eq. (3.32) with the forcing term $f(\tau + t)$. That is $S(\tau)Gf = GS(\tau)f$, i.e., G commutes with the translation group $(S(t))_{t\in\mathbb{R}}$.

Let $\lambda \in \Lambda$. We consider the function $f(t) := ae^{i\lambda t}$, $t \in \mathbb{R}$ with $0 \neq a \in \mathbb{X}$. By the above remark on the commutativeness of G and $(S(t))_{t\in\mathbb{R}}$, we have

$$\begin{aligned}\frac{dGf}{dt} &= \lim_{\tau\to 0} \frac{S(\tau)Gf - Gf}{\tau} \\ &= \lim_{\tau\to 0} G\frac{S(\tau)f - f}{\tau} \\ &= G\lim_{\tau\to 0} \frac{S(\tau)f - f}{\tau} \\ &= i\lambda Gf.\end{aligned}$$

This shows that $u(t) := [Gf](t) = be^{i\lambda t}$. Substituting this expression into the equation

$$u(t) = T(t-s)u(s) + \int_s^t T(t-s)f(s)ds,$$

we have

$$b = R(h)b + \int_0^h R(h-\xi)ad\xi,$$

where $R(h) = e^{-i\lambda h}T(h)$, $h = t-s$. Obviously, $(R(h))_{h\geq 0}$ is a C_0-semigroup with generator $A_\lambda = A - i\lambda I$. Then from the general theory of C_0-semigroup the above expression yields that

$$b \in D(A_\lambda) = D(A), \quad (A - i\lambda I)b = a.$$

It may be noted that b is unique. By the arbitrary nature of a and $\lambda \in \Lambda$ it follows that $\lambda \in \rho(A)$. This proves the theorem. □

We are now in a position to formulate a main result of this section:

Theorem 3.5. Λ *be a closed nonemty subset of the real line. Moreover let*

$$i\Lambda \cap \sigma(A) = \oslash. \tag{3.35}$$

Then for every $f \in \Lambda(\mathbb{X})$ Eq.(3.32) has a unique (classical) bounded solution in $\Lambda(\mathbb{X})$ provided one of the following conditions holds

(i) either Λ is compact, or
(ii) the operator A is bounded on $\mathbb{X}$.

In particular, the subspace $\Lambda(\mathbb{X})$ is admissible for Eq.(3.32) in both cases.

Proof. i) First of all by assumption, it is seen that the operator $\mathcal{D}_{\Lambda(\mathbb{X})}$ is bounded (see e.g. p. 88 in [Levitan and Zhikov (58)]). Since $\Lambda(\mathbb{X})$ satisfies also condition H1, by Lemma 3.7,

$$\sigma(\mathcal{A}_{\Lambda(\mathbb{X})}) \subset \sigma(A). \tag{3.36}$$

Thus, applying Theorem 1.11 to the pair of operators $\mathcal{D}_{\Lambda(\mathbb{X})}$ and $\mathcal{A}_{\Lambda(\mathbb{X})}$ we get the assertion of the theorem. we get the required.
ii) The second case can be proved in the same manner. □

We now consider a more general case where the operators $\mathcal{A}$ and $\mathcal{D}$ satisfy condition P

Theorem 3.6. *Let $(A+\alpha)$ be of class $\Sigma(\theta+\pi/2, R)$ for some real α and $\mathcal{M}$ be a translation - invariant subspace of $BUC(\mathbb{R},\mathbb{X})$. Moreover, let $\sigma(A) \cap \sigma(\mathcal{D}_{\mathcal{M}}) = \oslash$. Then the following assertions hold true:*

- *(i) If $\mathcal{M}$ satisfies condition H1, then $\mathcal{M}$ is weakly admissible for (3.32).*
- *(ii) If $\mathcal{M}$ satisfies condition H2 and A is the generator of a C_0-semigroup, then $\mathcal{M}$ is admissible, weakly admissible and mildly admissible for (3.32).*
- *(iii) If $\mathcal{M} \subset AAP(\mathbb{X})$ satisfies condition H1 and A is the generator of a C_0-semigroup, then $\mathcal{M}$ is admissible, weakly admissible and mildly admissible for (3.32).*

Proof. Note that under the theorem's assumption the operators $\mathcal{A}+\alpha$ and $\mathcal{D}$ satisfy condition P for some real α. In fact, we can check only that

$$\sup_{\lambda\in\Sigma(\pi/2-\varepsilon,R)} \|\lambda R(\lambda,\mathcal{D}_{\mathcal{M}})\| < \infty,$$

where $0<\varepsilon<\pi/2$. Since $\lambda \in \Sigma(\pi/2-\varepsilon, R)$ with $0<\varepsilon<\pi/2$

$$\begin{aligned}\|\lambda R(\lambda,\mathcal{D}_{\mathcal{M}})f\| &= |\lambda|\|\int_0^\infty e^{-\lambda t} f(\cdot+t)dt\| \\ &\le |\lambda|\int_0^\infty e^{-Re\lambda t}dt\|f\| \\ &\le \frac{|\lambda|}{Re\lambda}\|f\| \\ &\le M\|f\|,\end{aligned}$$

where M is a constant independent of f. Thus, by Theorem 1.11,

$$\begin{aligned}\sigma(\overline{\mathcal{D}_{\mathcal{M}}-\mathcal{A}_{\mathcal{M}}})^A - \alpha = \sigma(\overline{\mathcal{D}_{\mathcal{M}}-\mathcal{A}_{\mathcal{M}}}^A - \alpha) &= \sigma(\overline{\mathcal{D}_{\mathcal{M}}-(\mathcal{A}_{\mathcal{M}}+\alpha)})^A \\ &\subset \sigma(\mathcal{D}_{\mathcal{M}}) - \sigma(\mathcal{A}_{\mathcal{M}}+\alpha) \\ &\subset \sigma(\mathcal{D}_{\mathcal{M}}) - \sigma(\mathcal{A}_{\mathcal{M}}) - \alpha.\end{aligned}$$

Hence

$$\sigma(\overline{\mathcal{D}_{\mathcal{M}} - \mathcal{A}_{\mathcal{M}}}^{A}) \subset \sigma(\mathcal{D}_{\mathcal{M}}) - \sigma(\mathcal{A}). \tag{3.37}$$

By assumption and by Lemma 3.7 since $\sigma(\mathcal{D}_{\mathcal{M}}) \cap \sigma(A) = \oslash$ we have $\sigma(\mathcal{D}_{\mathcal{M}}) \cap \sigma(\mathcal{A}) = \oslash$. From (3.37) and this argument we get

$$0 \notin \sigma(\overline{\mathcal{D}_{\mathcal{M}} - \mathcal{A}_{\mathcal{M}}}^{A}).$$

Hence, this implies in particular the weak admissibility of the function space $\mathcal{M}$ for Eq.(3.32) proving i). Now in addition suppose that A generates a strongly continuous semigroup. Then ii) and iii) are immediate consequences of Corollary 3.6 and i). □

3.2.4 *Higher Order Differential Equations*

In this subsection we consider the admissibility of the function space $\mathcal{M} \cap \Lambda(\mathbb{X})$ where $\mathcal{M}$ is assumed to satisfy condition H1 and Λ is a closed subset of the real line for the equation

$$\frac{d^n u}{dt^n} = Au + f(t), \tag{3.38}$$

where n is a natural number. To this end, we first study the operator $d^n u/dt^n := \mathcal{D}^n$ on $\mathcal{M} \cap \Lambda(\mathbb{X})$.

Proposition 3.3. *With the above notation the following assertions hold true:*

(i)

$$\sigma(\mathcal{D}^n_{\mathcal{M}\cap\Lambda(\mathbb{X})}) \subset (i\Lambda)^n.$$

(ii)

$$\sigma(\mathcal{D}^n_{\Lambda(\mathbb{X})}) = (i\Lambda)^n.$$

Proof. We associate with the equation

$$\frac{d^n u}{dt^n} = \mu u + f(t), f \in \mathcal{M} \cap \Lambda(\mathbb{X})$$

the following first order equation

$$\begin{cases} x_1' = x_2 \\ x_2' = x_3, \\ \dots \\ x_n' = \mu x_1 + f(t) \end{cases} \qquad f \in \mathcal{M} \cap \Lambda(\mathbb{X}). \tag{3.39}$$

Obviously, the unique solvability of these equations in the function space $\mathcal{M} \cap \Lambda(\mathbb{X})$ are equivalent. On the other hand, by Theorem 3.5 for every $f \in \mathcal{M} \cap \Lambda(\mathbb{X})$ Eq.(3.39) has a unique (classical) solution $x(\cdot) \in \mathcal{M} \cap \Lambda(\mathbb{X}), x = (x_1, \cdots, x_n)^T$ if

$$i\Lambda \cap \sigma(I(\mu)) = \oslash,$$

where $I(\mu)$ denotes the operator matrix associated with Eq.(3.39). A simple computation shows that $\sigma(I(\mu))$ consists of all solutions to the equation $t^n - \mu = 0$. Thus,

$$\sigma(\mathcal{D}^n_{\mathcal{M}\cap\Lambda(\mathbb{X})}) \subset \{\mu \in \mathbf{C} : \mu = (i\lambda)^n \text{for some} \lambda \in \Lambda\}.$$

Hence i) is proved. On the other hand, let $\mu \in \Lambda$. Then $g(\cdot) := xe^{i\mu\cdot} \in \Lambda(\mathbb{X})$. Obviously, $\mathcal{D}^n_{\Lambda(\mathbb{X})} g = (i\mu)^n g$ and thus, $(i\mu)^n \in \sigma(\mathcal{D}^n_{\Lambda(\mathbb{X})})$. Hence, ii) is proved. □

To proceed we make a definition

Definition 3.8. The definition of *admissibility for the first order equations is naturally extended to higher order equations.*

Observe that $(i\Lambda)^n$ is compact if Λ is compact.

Theorem 3.7. *Let Λ be a compact subset of the real line and $\mathcal{M}$ be a translation invariant subspace of $BUC(\mathbb{R}, \mathbb{X})$ satisfying condition H1. Moreover, let A be any closed operator in $\mathbb{X}$ such that $\sigma(A) \cap (i\Lambda)^n = \oslash$. Then for every $f \in \mathcal{M} \cap \Lambda(\mathbb{X})$ there exists a unique (classical) solution $u_f \in \mathcal{M} \cap \Lambda(\mathbb{X})$ of Eq.(3.38). In particular, $\mathcal{M} \cap \Lambda(\mathbb{X})$ is admissible for Eq.(3.38).*

Proof. The theorem is an immediate consequence of Theorem 3.5 and the above computation of the spectrum of $\mathcal{D}^n$. □

We recall the following notion.

Definition 3.9. By a *mild solution of Eq.(3.38)* we understand a bounded uniformly continuous function $u : \mathbb{R} \to \mathbb{X}$ such that

$$\int_0^t dt_1 \int_0^{t_1} dt_2 ... \int_0^{t_{n-1}} u(s)ds \in D(A)$$

and

$$u(t) = x_0 + tx_1 + ...t^{n-1}x_{n-1} + A\int_0^t dt_1 \int_0^{t_1} dt_2 ... \int_0^{t_{n-1}} u(s)ds$$
$$+ \int_0^t dt_1 \int_0^{t_1} dt_2 ... \int_0^{t_{n-1}} f(s)ds \quad (t \in \mathbb{R})$$

for some fixed $x_0, x_1, ..., x_{n-1} \in \mathbb{X}$. For $u \in BUC(\mathbb{R}, \mathbb{X})$ we say that u is a *classical solution* to Eq.(3.38) if $u(t) \in D(A), \forall t \in \mathbb{R}$ and the n-th derivative of u (denoted by $u^{(n)}$) exists as an element of $BUC(\mathbb{R}, \mathbb{X})$ such that Eq.(3.38) holds for all $t \in \mathbb{R}$.

Remark 3.4. Obviously, a classical solution is also a mild solution. In case $n = 1$ if A generates a strongly continuous semigroup the above definition of mild solution on $\mathbb{R}$ coincides with the usual notion which we have used so far. In fact we have:

Lemma 3.10. *Let A be the generator of a C_0-semigroup and u satisfy $\int_0^t u(s)ds \in D(A)$, $\forall t$ such that*

$$u(t) = u(0) + A \int_0^t u(s)ds + \int_0^t f(s)ds. \tag{3.40}$$

Then u satisfies

$$u(t) = T(t-s)u(s) + \int_s^t T(t-\xi)f(\xi)d\xi, \forall t \geq s \tag{3.41}$$

where $T(t) = e^{tA}$. Conversely, if u satisfies Eq.(3.41), then $\int_0^t u(s)ds \in D(A)$, $\forall t$ and u satisfies Eq.(3.40).

Proof. Suppose that u is a solution to Eq.(3.40). Then, we will show that it is also a solution to Eq.(3.41). In fact, without loss of generality we verify that u satisfies Eq.(3.41) for $s = 0$. To this purpose let us define the function

$$w(t) = T(t)u(0) + \int_0^t T(t-\xi)f(\xi)d\xi, t \geq 0.$$

We now show that w satisfies Eq.(3.40) for $t \geq 0$ as well. In fact, using the following facts from semigroup theory

$$T(t)x - x = A \int_0^t T(s)xds, \forall x \in \mathbb{X},$$

and the following which can be verified directly by definition

$$\begin{aligned} A \int_\eta^t T(s-\eta)f(\eta)ds &= \lim_{h \downarrow 0}(1/h)(T(h) - I) \int_\eta^t T(s-\eta)f(\eta)ds \\ &= T(t-\eta)f(\eta) - f(\eta) \end{aligned}$$

we have

$$A \int_0^t w(s)ds = T(t)u(0) - u(0) + A \int_0^t \int_0^s T(s-\eta)f(\eta)d\eta.$$

By a change of order of integrating we get

$$A\int_0^t\int_0^s T(s-\eta)f(\eta)d\eta = A\int_0^t d\eta\int_\eta^t T(s-\eta)f(\eta)ds.$$

By the above facts we have

$$\int_0^t d\eta A\int_\eta^t T(s-\eta)f(\eta)ds = \int_0^t T(t-\eta)f(\eta)d\eta - \int_0^t f(\eta)d\eta.$$

This shows that w satisfies Eq.(3.40) for all $t \geq 0$. Define $g(t) = w(t) - u(t)$. Obviously,

$$g(t) = A\int_0^t g(s)ds, \forall t \geq 0.$$

Since A generates a strongly continuous semigroup the Cauchy problem

$$x' = Ax, x(0) = 0 \in D(A)$$

has a unique solution zero. Hence, $u(t) = w(t), \forall t \geq 0$, i.e. $u(t)$ satisfies Eq.(3.41) for all $t \geq 0$.

By reversing the above argument we can easily show the converse. Hence, the lemma is proved. □

Lemma 3.11. *Let A be a closed operator and u be a mild solution of Eq.(3.38) and $\phi \in L^1(\mathbb{R})$ such that its Fourier transform has compact support. Then $\phi * u$ is a classical solution to Eq.(3.38) with forcing term $\phi * f$.*

Proof. Let us define

$$U_1(t) = \int_0^t u(s)ds, \quad F_1(t) = \int_0^t f(s)ds, \ t \in \mathbb{R},$$

$$U_k(t) = \int_0^t U_{k-1}(s)ds, \quad F_k(t) = \int_0^t F_{k-1}(s)ds, \ t \in \mathbb{R}, k \in \mathbf{N}.$$

Then, by definition, we have

$$u(t) = P_n(t) + A(U_n(t)) + F_n(t), t \in \mathbb{R},$$

where P_n is a polynomial of order of $n-1$. From the closedness of A, we have

$$u * \phi(t) = P_n * \phi(t) + A(U_n * \phi(t)) + F_n * \phi(t), \ t \in \mathbb{R}.$$

Since the Fourier transform ϕ has compact support all convolutions above are infinitely differentiable. From the closedness of A we have that $(U_n * \phi)^{(k)}(t) \in D(A), t \in \mathbb{R}$ and

$$A((U_n * \phi)^{(k)}(t)) = \frac{d^k}{dt^k}A(U_n * \phi(t)), \ t \in \mathbb{R}.$$

Set $V_n(t) = U_n*\phi(t)$. Since $U_n^{(k)}(t) = U_{n-k}(t), k = 0,1,2\cdots,n$ and $U_0(t) = u(t)$ we have $V^{(k)}(t) = U_{n-k}*\phi(t), k = 0,1,2,\cdots,n$ and $V_n^{(n)}(t) = u*\phi(t)$. Hence,

$$\begin{aligned} U_n * \phi(t) &= V_n(t) \\ &= \sum_{k=0}^{n-1} \frac{t^k}{k!} U_{n-k} * \phi(0) + \frac{1}{(n-1)!}\int_0^t (t-s)^{n-1} u * \phi(s)ds. \end{aligned}$$

Since $U_n * \phi(t) \in D(A), U_{n-k} * \phi(0) \in D(A), k = 1,2,\cdots,n-1$ it follows that the integral above belongs also to $D(A)$. Furthermore, we can check that

$$\begin{aligned} u * \phi(t) = P_n * \phi(t) + Q_n(t) + A\left(\frac{1}{(n-1)!}\int_0^t (t-s)^{n-1} u * \phi(s)ds\right) \\ + \frac{1}{(n-1)!}\int_0^t (t-s)^{n-1} f * \phi(s)ds, \end{aligned}$$

where P_n, Q_n are polynomials of order of $n-1$ which appears when one expands $A(U_n * \phi(t))$ and $F_n * \phi(t)$, respectively. Now, since all functions in the above expression are infinitely differentiable, P_n, Q_n are polynomials of order of $n-1$ and A is closed we can differentiate the expression to get

$$\frac{d^n}{dt^n}(u*\phi)(t) = A(u*\phi(t)) + f*\phi(t), \ \forall t \in \mathbb{R}.$$

This proves the lemma. □

We now recall the notion of Λ-class of functions.

Definition 3.10. A translation invariant subspace $\mathcal{F} \subset BUC(\mathbb{R}, \mathbb{X})$ is said to be a Λ-*class* if and only if it satisfies

(1) $\mathcal{F}$ is a closed subspace of $BUC(\mathbb{R}, \mathbb{X})$;
(2) $\mathcal{F}$ contains all constant functions;
(3) $\mathcal{F}$ satisfies condition H1;
(4) $\mathcal{F}$ is invariant by multiplication by $e^{i\xi\cdot}, \forall \xi \in \mathbb{R}$.

Let $\mathcal{F}$ be a Λ-class and u be in $BUC(\mathbb{R}, \mathbb{X})$. Then, by definition

$$\begin{aligned} sp_{\mathcal{F}}(u) := \{\xi \in \mathbb{R} : \forall \varepsilon > 0 \exists f \in L^1(\mathbb{R}) \\ \text{such that } supp\mathcal{F}f \subset (\xi-\varepsilon, \xi+\varepsilon) \text{ and } f*u \notin \mathcal{F}\}. \end{aligned}$$

Lemma 3.12. *If $f \in \mathcal{F}$, where $\mathcal{F}$ is a Λ-class, then $\psi * f \in \mathcal{F}$, $\forall \psi \in L^1(\mathbb{R})$ such that the Fourier transform of ψ has compact support.*

Proof. For the proof we refer the reader to p.60 in [Basit (12)]. □

Hence, Theorem 3.7 yields the following:

Theorem 3.8. *Let $\mathcal{F}$ be a Λ-class, A be a closed linear operator with non-empty resolvent set. Then for any mild solution u to Eq.(3.38) with $f \in \mathcal{F}$,*

$$sp_{\mathcal{F}}(u) \subset \{\lambda \in \mathbb{R} : (i\lambda)^n \in \sigma(A)\}. \tag{3.42}$$

Proof. Let $\lambda_0 \in \mathbb{R}$ such that $(i\lambda_0)^n \notin \sigma(A)$. Then, since $\sigma(A)$ is closed there is a positive number δ such that for all $\lambda \in (\lambda_0 - 2\delta, \lambda_0 + 2\delta)$ we have $(i\lambda)^n \notin \sigma(A)$. Let us define $\Lambda := [\lambda_0 - \delta, \lambda_0 + \delta]$. Then by Theorem 3.7 for every $y \in \Lambda(\mathbb{X}) \cap \mathcal{F}$ there is a unique (classical) solution $x \in \Lambda(\mathbb{X}) \cap \mathcal{F}$. Let $\psi \in L^1(\mathbb{R})$ such that $supp\mathcal{F}\psi \subset \Lambda$. Put $v := \psi * u, g := \psi * f$. Then, by Lemma 3.12 $g \in \mathcal{F}$ and by Proposition 2.5 in [Basit (12)] $sp_{\mathcal{F}}(g) \subset supp\mathcal{F}\psi \cap sp_{\mathcal{F}}(f) \subset \Lambda$. Thus $g \in \Lambda(\mathbb{X}) \cap \mathcal{F}$. Since $sp_{\mathcal{F}}(v) \subset \Lambda$ by Theorem 3.7 we see that Eq.(3.38) has a unique solution in $\Lambda(\mathbb{X})$ which should be v. Moreover, applying again Theorem 3.7 we can see that the function v should belong to $\Lambda(\mathbb{X}) \cap \mathcal{F}$. We have in fact proved that $\lambda_0 \notin sp_{\mathcal{F}}(u)$. Hence the assertion of the theorem has been proved. □

In a standard manner we get the following:

Corollary 3.7. *Let $\mathcal{F}$ be a Λ-class, $\sigma(A) \cap (i\mathbb{R})^n$ be countable. Moreover, let u be such a mild solution to Eq.(3.38) that*

$$\lim_{t\to\infty} \frac{1}{t} \int_0^t e^{-i\lambda s} u(x+s)ds$$

exists for every $\lambda \in sp_{\mathcal{F}}(u)$ uniformly with respect to $x \in \mathbb{R}$. Then $u \in \mathcal{F}$.

Proof. The corollary is an immediate consequence of Theorem 1.18 and Theorem 3.8. □

In particular, we can take $\mathcal{F} = AP(\mathbb{X}), AAP(\mathbb{X})$ and get spectral criteria for almost periodicity and asymptotic almost periodicity for solutions to the higher order equations (3.38).

Next, we consider the admissibility of a given translation invariant closed subspace $\mathcal{M}$ for the higher order equation (3.38). Since the geometric properties of the set $(i\mathbb{R})^n$ play an important role, we consider here only the case $n = 2$, i.e., the following equation

$$\frac{d^2u}{dt^2} = Au + f(t). \tag{3.43}$$

It turns out that for higher order equations conditions on A are much weaker than for the first order ones. Indeed, we have

Theorem 3.9. *Let A be a linear operator on $\mathbb{X}$ such that there are positive constants R, θ and*

$$\Sigma(\theta, R) \subset \rho(A) \quad \textit{and} \quad \sup_{\lambda \in \Sigma(\theta, R)} |\lambda| \|R(\lambda, A)\| < \infty.$$

Furthermore, let $\mathcal{M}$ be a translation invariant closed subspace of $BUC(\mathbb{R}, \mathbb{X})$ which satisfies condition H1 such that

$$\sigma(\mathcal{D}^2_{\mathcal{M}}) \cap \sigma(A) = \oslash.$$

Then $\mathcal{M}$ is admissible for the second order equation (3.43).

Proof. We will apply Theorem 1.11 to the pair of linear operators $\mathcal{D}^2_{\mathcal{M}}$ and $\mathcal{A}_{\mathcal{M}}$. To this end, by Proposition 3.3 we observe that

$$\sigma(\mathcal{D}^2_{\mathcal{M}}) \subset (i\mathbb{R})^2 = (-\infty, 0].$$

On the other hand, for $0 < \varepsilon < \theta$ we can show that there is a constant M such that the following estimate holds

$$\|R(\lambda, \mathcal{D}^2_{\mathcal{M}})\| \leq \frac{M}{|\lambda|}, \forall \lambda \neq 0, |arg(\lambda) - \pi| < \varepsilon.$$

In fact, this follows immediately from well known facts in Chapter 2 in [Daleckii and Krein (24)]. To make it more clear, we consider the first order equation of the form (3.39) for the case $n = 2$. For every $\lambda \in \rho(\mathcal{D}^2_{\mathcal{M}})$ the associated equation has an exponential dichotomy and its Green function is nothing but $R(\lambda, \mathcal{D}^2_{\mathcal{M}})$. Furthermore, since $\mathcal{M}$ is translation invariant note that $D(\mathcal{D}^2_{\mathcal{M}})$ is dense in $\mathcal{M}$. Thus, applying Theorem 1.11 to the pair of operators $\mathcal{D}^2_{\mathcal{M}}, \mathcal{A}_{\mathcal{M}}$ we have

$$0 \in \rho(\overline{\mathcal{D}^2_{\mathcal{M}} - \mathcal{A}_{\mathcal{M}}}).$$

It remains to show that for every $f \in \mathcal{M}_0 := D(\mathcal{D}^2_{\mathcal{M}})$ there is a unique classical solution u on $\mathbb{R}$. In fact, denoting

$$G := (\overline{\mathcal{D}^2_{\mathcal{M}} - \mathcal{A}_{\mathcal{M}}})^{-1},$$

we can easily see that since $\mathcal{D}^2_{\mathcal{M}}, \mathcal{A}_{\mathcal{M}}$ commute with $\mathcal{D}^2_{\mathcal{M}}$, so does G. By definition, for $\lambda \in \rho(\mathcal{D}^2_{\mathcal{M}})$, since G is bounded on $\mathcal{M}$

$$GR(\lambda, \mathcal{D}^2_{\mathcal{M}}) = R(\lambda, \mathcal{D}^2_{\mathcal{M}})G.$$

Hence there is $g \in \mathcal{M}$ such that $f = R(\lambda, \mathcal{D}^2_{\mathcal{M}})g$. Thus, by the above equality $Gf = R(\lambda, \mathcal{D}^2_{\mathcal{M}})Gg \in D(\mathcal{D}^2_{\mathcal{M}})$. This shows the admissibility of $\mathcal{M}$ for Eq.(3.43). □

3.2.5 *Abstract Functional Differential Equations*

This subsection will be devoted to some generalization of the method discussed in the previous ones for functional differential equations of the form

$$\frac{dx(t)}{dt} = Ax(t) + [\mathcal{B}x](t) + f(t), \forall t \in \mathbb{R}, \tag{3.44}$$

where the operator A is a linear operator on $\mathbb{X}$ and $\mathcal{B}$ is assumed to be an autonomous functional operator.

We first precise the notion of *autonomousness* for functional operators $\mathcal{B}$:

Definition 3.11. Let $\mathcal{B}$ be an operator, everywhere defined and bounded on $BUC(\mathbb{R}, \mathbb{X})$ into itself. $\mathcal{B}$ is said to be an *autonomous functional operator* if for every $\phi \in BUC(\mathbb{R}, \mathbb{X})$

$$S(\tau)\mathcal{B}\phi = \mathcal{B}S(\tau)\phi, \forall \tau \in \mathbb{R},$$

where $(S(\tau))_{\tau \in \mathbb{R}}$ is the translation group $S(\tau)x(\cdot) := x(\tau + \cdot)$ in $BUC(\mathbb{R}, \mathbb{X})$.

In connection with autonomous functional operators we will consider closed translation invariant subspaces $\mathcal{M} \subset BUC(\mathbb{R}, \mathbb{X})$ which satisfy condition H3. Recall that if $\mathcal{B}$ is an autonomous functional operator and $\mathcal{M}$ satisfies condition H3, then by definition, $\mathcal{M}$ is left invariant under $\mathcal{B}$.

Definition 3.12. Let A be the generator of a C_0-semigroup and $\mathcal{B}$ be an autonomous functional operator. A function u on $\mathbb{R}$ is said to be a *mild solution* of Eq.(3.44) on $\mathbb{R}$ if

$$u(t) = e^{(t-s)A}u(s) + \int_s^t e^{(t-\xi)A}[(Bu)(\xi) + f(\xi)]d\xi, \forall t \geq s.$$

As we have defined the notion of mild solutions it is natural to extend the notion of mild admissibility for Eq.(3.44) in the case where the operator A generates a strongly continuous semigroup. It is interesting to note that in this case because of the arbitrary nature of an autonomous functional operator $\mathcal{B}$ nothing can be said on the "well posedness" of Eq.(3.44). We refer the reader to Chapter 1 for particular cases of "finite delay" and "infinite delay" in which Eq.(3.44) is well posed.) However, as shown below we can extend our approach to this case. Now we formulate the main result for this subsection.

Theorem 3.10. *Let A be the infinitesimal generator of an analytic strongly continuous semigroup, $\mathcal{B}$ be an autonomous functional operator on the function space $BUC(\mathbb{R}, \mathbb{X})$ and $\mathcal{M}$ be a closed translation invariant subspace of $AAP(\mathbb{X})$ which satisfies condition H3. Moreover, assume that*

$$\sigma(\mathcal{D}_{\mathcal{M}}) \cap \sigma(\mathcal{A} + \mathcal{B}) = \oslash.$$

Then $\mathcal{M}$ is mildly admissible for Eq. (3.44), i.e., for every $f \in \mathcal{M}$ there is a unique mild solution $u_f \in \mathcal{M}$ of Eq.(3.44).

Proof. Since $\mathcal{M}$ satisfies condition H3, for every $f \in \mathcal{M}$ we have $\mathcal{B}f \in \mathcal{M}$. Thus,

$$\begin{aligned} D((\mathcal{A}+\mathcal{B})_{\mathcal{M}}) &= \{f \in \mathcal{M} : Af(\cdot) + \mathcal{B}f \in \mathcal{M}\} \\ &= \{f \in \mathcal{M} : Af(\cdot) \in \mathcal{M}\} \\ &= D(\mathcal{A}_{\mathcal{M}}). \end{aligned}$$

Hence

$$(\mathcal{A}+\mathcal{B})_{\mathcal{M}} = \mathcal{A}_{\mathcal{M}} + \mathcal{B}_{\mathcal{M}}.$$

As $\mathcal{M}$ satisfies condition H3 it satisfies condition H1 as well. Thus, by Lemma 3.7,

$$\sigma(\mathcal{A}_{\mathcal{M}}) \subset \sigma(\mathcal{A}) \subset \sigma(A)$$

and

$$\|R(\lambda, \mathcal{A}_{\mathcal{M}})\| \leq \|R(\lambda, A)\|, \forall \lambda \in \rho(A).$$

Since $\mathcal{B}$ is bounded $\mathcal{D}_{\mathcal{M}}$ and $(\mathcal{A}+\mathcal{B})_{\mathcal{M}} = \mathcal{A}_{\mathcal{M}} + \mathcal{B}_{\mathcal{M}}$ satisfy condition P. From Lemma 2 and the remarks follows in [Naito and Minh (74)] it may be seen that $\mathcal{A}_{\mathcal{M}}$ is the infinitesimal generator of the strongly continuous semigroup $(T(t))_{t\geq 0}$

$$T(t)f(\xi) := e^{tA}f(\xi), \forall f \in \mathcal{M}, \xi \in \mathbb{R}.$$

Hence $D((\mathcal{A}+\mathcal{B})_{\mathcal{M}}) = D(\mathcal{A}_{\mathcal{M}})$ is dense everywhere in $\mathcal{M}$. It may be noted that $R(\lambda, \mathcal{A}+\mathcal{B})$ commutes with the translation group. Since $\mathcal{M}$ satisfies condition H3 we can easily show that

$$\sigma((\mathcal{A}+\mathcal{B})_{\mathcal{M}}) \subset \sigma(\mathcal{A}+\mathcal{B}).$$

Applying Theorem 1.11 we get

$$\sigma(\overline{\mathcal{D}_{\mathcal{M}} - (\mathcal{A}+\mathcal{B})_{\mathcal{M}}}) \subset \sigma(\mathcal{D}_{\mathcal{M}}) - \sigma((\mathcal{A}+\mathcal{B})_{\mathcal{M}}).$$

Hence

$$0 \in \rho(\overline{\mathcal{D}_{\mathcal{M}} - (\mathcal{A}+\mathcal{B})_{\mathcal{M}}}).$$

On the other hand, since $\mathcal{B}_{\mathcal{M}}$ is bounded on $\mathcal{M}$

$$\begin{aligned}\overline{\mathcal{D}_{\mathcal{M}} - (\mathcal{A}+\mathcal{B})_{\mathcal{M}}} &= \overline{\mathcal{D}_{\mathcal{M}} - \mathcal{A}_{\mathcal{M}}} - \mathcal{B}_{\mathcal{M}} \\ &= L_{\mathcal{M}} - \mathcal{B}_{\mathcal{M}}\end{aligned}$$

we have

$$0 \in \rho(L_{\mathcal{M}} - \mathcal{B}_{\mathcal{M}}). \tag{3.45}$$

If $u, f \in \mathcal{M}$ such that $(L_{\mathcal{M}} - \mathcal{B}_{\mathcal{M}})u = f$, then

$$L_{\mathcal{M}}u = \mathcal{B}_{\mathcal{M}}u + f.$$

By definition of the operator $L_{\mathcal{M}}$, this is equivalent to the following

$$u(t) = e^{(t-s)A}u(s) + \int_s^t e^{(t-\xi)A}[(\mathcal{B}_{\mathcal{M}}u)(\xi) + f(\xi)]d\xi, \forall t \geq s,$$

i.e., u is a mild solution to Eq.(3.44). Thus (3.45) shows that $\mathcal{M}$ is mildly admissible for Eq.(3.44). □

Remark 3.5. Sometime it is convenient to re-state Theorem 3.10 in other form than that made above. In fact, in practice we may encounter difficulty in computing the spectrum $\sigma(\mathcal{A}+\mathcal{B})$. Hence, alternatively, we may consider $\mathcal{D}-\mathcal{A}-\mathcal{B}$ as a sum of two commuting operators $\mathcal{D}-\mathcal{B}$ and $\mathcal{A}$ if $\mathcal{B}$ commutes with $\mathcal{A}$. In subsection 3.4 we again consider this situation.

3.2.6 *Examples and Applications*

In this subsection we will present several examples and applications and discuss the relation between our results and the previous ones.

As typical examples of the function spaces $\Lambda(\mathbb{X})$, where Λ is a closed subset of the real line we will take the following ones:

Example 1 The space of all $\mathbb{X}$ valued continuous τ-periodic functions $\mathcal{P}(\tau)$. In this case $\Lambda = \{2k\pi/\tau, k \in \mathbf{Z}\}$.

Example 2 Let Λ be a discrete subset of $\mathbb{R}$. Then $\Lambda(\mathbb{X})$ will consists of almost periodic functions.

Example 3 Let Λ be a countable subset of $\mathbb{R}$. Then $\Lambda(\mathbb{X})$ will consists of almost periodic functions if in addition one assumes that $\mathbb{X}$ does not contain any subspace which is isomorphic to the space c_0 (see [Levitan and Zhikov (58)]).

Below we will revisit one of the main results of [Langenhop (53)] to show how our method fits in the problem considered in [Langenhop (53)]. Moreover, our method can be easily extended to the infinite dimensional case.

Example 4 (cf. [Langenhop (53)]) Consider the following ordinary functional differential equation

$$x'(t) = \int_0^\infty [dE(s)]x(t-s) + f(t), x \in \mathbf{C}^n, t \in \mathbb{R}, \tag{3.46}$$

where E is an $n \times n$ matrix function with elements in $\mathbf{C}$, f is a $\mathbf{C}^n$-valued almost periodic function. In addition, we assume that E is continuous from the left and of bounded total variation on $[0, \infty)$, i.e.

$$0 < \gamma = \int_0^\infty |dE(s)| < \infty.$$

As is well known for every $f \in AP(\mathbf{C}^n)$ there is a corresponding Fourier series

$$\sum_{k=0}^\infty a_k e^{i\lambda_k t}.$$

We define

$$\mathcal{A}_q^0 := \{f \in AP(\mathbf{C}^n) : a_0 = 0, |\lambda_k| \geq q, k = 1, 2, \cdots\}$$

and $\mathcal{A}_q = \mathcal{A}_q^0 + V_c$, where V_c is the set of all $\mathbf{C}^n$-valued constant functions. Now we define our operator

$$\mathcal{B}u(t) := \int_0^\infty [dE(s)]u(t-s), t \in \mathbb{R}, u \in AP(\mathbf{C}^n).$$

Obviously, $\mathcal{B}$ is an autonomous functional operator with $\|\mathcal{B}\| \leq \gamma$. If we define $\Lambda := \{\eta \in \mathbb{R} : |\eta| \geq q\}$, then $\mathcal{A}_q^0 = AP(\mathbf{C}^n) \cap \Lambda(\mathbf{C}^n)$. Now we prove the following:

Assertion 1 *Under the above notations and assumptions Eq.(3.46) has a unique almost periodic solution $x_f \in \mathcal{A}_q^0$ for every $f \in \mathcal{A}_q^0$ if $\gamma < q$.*

Proof. In fact, by assumption it is obvious that the spectral radius $r_\sigma(\mathcal{B}) < \gamma$. Hence, $i\Lambda \cap \sigma(\mathcal{B}) = \oslash$. □

If in addition we assume that

$$M := \int_0^\infty dE(s) \tag{3.47}$$

is a nonsigular matrix, then Assertion 1 implies the following:

Assertion 2 *Under Assertion 1's assumptions and the nonsingularity of the matrix (3.47) there exists a unique solution $x_f \in \mathcal{A}_q$ to Eq.(3.46) for every $f \in \mathcal{A}_q$.*

Proof. In this case the operator $d/dt - \mathcal{B}$ is a direct sum of two invertible operators in $\mathcal{A}_q^0$ and V_c. □

Remark 3.6. In Section 3 in [Langenhop (53)] Assertion 2 has been proved with a little stronger assumption, namely, $\gamma\delta < q$, where $\delta \geq 1$ is an "absolute constant" (in terminology of p.401 in [Langenhop (53)]). The condition $\gamma < q$ of Assertion 2 becomes also necessary in many cases. To show this, we consider the case

$$\mathcal{B}u(t) = Bu(t+\tau), \forall t \in \mathbb{R}, u \in BUC(\mathbb{R}, \mathbb{X}),$$

where τ is a given constant, B is a matrix. Now suppose that there exists a unique solution $x_f \in \mathcal{A}_q$ to Eq.(3.46) for every $f \in \mathcal{A}_q$. Denoting $Gf := x_f$ we see that G is a bounded linear operator on $\mathcal{A}_q$. Moreover, since $\mathcal{B}$ commutes with translation group so does G , i.e., $\mathcal{D}Gf = G\mathcal{D}f, \forall f \in D(\mathcal{D})$. Taking $f := e^{i\lambda t}y$ we have $\mathcal{D}x_f = \mathcal{D}Gf = G\mathcal{D}f = \lambda Gf = \lambda x_f$. Hence, $x_f(t) = e^{i\lambda t}x$ for some x. Substituting this into Eq.(3.46) we get the assertion that given $|\lambda| \geq q$

for every $y \in \mathbf{C}^n$ there exists a unique $x \in \mathbf{C}^n$ such that

$$i\lambda x - e^{i\tau\lambda}Bx = y.$$

This shows that $i\lambda \in \rho(e^{i\tau\lambda}B) = e^{i\tau\lambda}\rho(B)$ and yields $\gamma < q$.

In the following example we will revisit a problem discussed in [Sljusarcuk (98)] with an unbounded A.

Example 5 Let us consider the equation

$$\frac{dx(t)}{dt} = Ax(t) + \sum_{k=1}^{N} B_k x(t+\tau_k) + f(t), t \in \mathbb{R}, \tag{3.48}$$

where A is the infinitesimal generator of an analytic C_0-semigroup, $B_k, k = 1, \cdots, N$ are bounded linear operators on $\mathbb{X}$ which are commutative with each other and A , $\tau_k, k = 1, \cdots, N$ are given reals and f is a bounded uniformly continuous function. We denote $\Lambda = sp(f)$.

Assertion 3 *Let Λ be bounded. Then if*

$$\sigma(A) \cap \cup_{\lambda\in\Lambda}\sigma(i\lambda - \sum_{k=1}^{N} B_k)e^{i\tau_k\lambda}) = \oslash,$$

Eq.(3.48) has a unique classical solution in $\Lambda(\mathbb{X})$.

Proof. If Λ is bounded , then $\mathcal{D}_{\Lambda(\mathbb{X})}$ is bounded. Hence, if

$$\sigma(\mathcal{D}_{\Lambda(\mathbb{X})} - \sum_{k=1}^{N} B_k e^{\tau_k \mathcal{D}_{\Lambda(\mathbb{X})}}) \cap \sigma(A) = \oslash$$

Eq.(3.48) has a unique classical solution in $\Lambda(\mathbb{X})$. In turn, using the estimates of spectra as in [Sljusarcuk (98)] we get

$$\sigma(\mathcal{D}_{\Lambda(\mathbb{X})} - \sum_{k=1}^{N} B_k e^{\tau_k \mathcal{D}_{\Lambda(\mathbb{X})}}) \subset \cup_{\lambda\in\Lambda}\sigma(i\lambda - \sum_{k=1}^{N} B_k e^{i\tau_k\lambda}).$$

□

Assertion 4 *Let f be almost periodic and $B_k, k = 1, \cdots, N$ be commutative with each other and A and*

$$i\Lambda \cap (\sigma(A) + \cup_{\lambda_k \in \overline{e^{i\tau_k\Lambda}}}\sigma(\sum_{k=1}^{N} B_k\lambda_k)) = \oslash.$$

Then Eq.(3.48) has a unique almost periodic mild solution in $\Lambda(\mathbb{X})$.

Proof. First using the Weak Spectral Mapping Theorem (see e.g. [Nagel (73)]) we have

$$\sigma(S(\tau_k)) = \overline{e^{i\tau_k\Lambda}}.$$

In view of Theorem 1 in [Sljusarcuk (98)], denoting the multiplication operator by B_k by also B_k for the sake of simplicity, we have

$$\sigma(\sum_{k=1}^{N} B_k S(\tau_k)) \subset \cup_{\lambda_k \in \overline{e^{i\tau_k\Lambda}}}\sigma(\sum_{k=1}^{N} B_k\lambda_k).$$

By the commutativeness assumption applying Theorem 3.10 and then Theorem 1.11 we get the conclusion of the assertion. □

In case A is the generator of a C_0-semigroup which is not necessarily analytic we can still apply Theorem 3.2 and the commutativeness of the operators $\mathcal{A}, \mathcal{B}$ as shown in the following example:

Example 6 Let A be the infinitesimal generator of a strongly continuous semigroup of linear operators on $\mathbb{X}$, $\mathcal{B}$ be an autonomous functional operator on $BUC(\mathbb{R},\mathbb{X})$ and $\mathcal{M}$ be a translation invariant subspace of $AAP(\mathbb{X})$. Moreover, we assume that $\mathcal{B}$ and $\mathcal{A}$ commute. The only difference between this example and the previous one is that the semigroup generated by A may not be analytic. However, we can find conditions for the admissibility of $\mathcal{M}$ by using evolution semigroup associated with A as in Theorem 3.2. In fact, in $\mathcal{M}$, since $\mathcal{B}_{\mathcal{M}}$ is bounded it generates the norm continuous semigroup $(B^h)_{h\geq 0}$. Hence, $\overline{-\mathcal{D}_{\mathcal{M}} + \mathcal{A}_{\mathcal{M}}} + \mathcal{B}_{\mathcal{M}}$ generates a strongly continuous semigroup $(T^h B^h)_{h\geq 0}$. Thus, in view of spectral inclusion of strongly continuous semigroups, this generator is invertible if $1 \notin \sigma(T^1 B^1)$. Using the commutativeness of the operators under consideration and the Weak Spectral Mapping Theorem for the translation group on $\mathcal{M}$ we have

$$\sigma(T^1 B^1) \subset \sigma(T^1).\sigma(B^1) \subset \overline{e^{-\mathcal{D}_{\mathcal{M}}}\sigma(e^A)}.\sigma(B^1). \tag{3.49}$$

Hence the following is obvious:

Assertion 5 *If*

$$1 \notin \overline{e^{-\mathcal{D}_{\mathcal{M}}}\sigma(e^A)}.\sigma(B^1),$$

then

$$\frac{dx(t)}{dt} = Ax(t) + [\mathcal{B}x](t) + f(t), \tag{3.50}$$

has a unique mild solution in $\mathcal{M}$ for every given $f \in \mathcal{M}$.

As an application suppose that we are given an almost periodic function f. Let $\mathcal{M} \subset AP(\mathbb{X})$ consisting of all functions g such that $sp(g) \subset sp(f)$. Then the above condition can be written as

$$1 \notin \overline{e^{-isp(f)}\sigma(e^A)}.e^{\sigma(\mathcal{B}_{\mathcal{M}})} \tag{3.51}$$

which implies the existence of an almost periodic mild solution to Eq.(3.49). To illustrate the usefulness of (3.51) we consider the following case of Eq.(3.50)

$$\frac{dx(t)}{dt} = Ax(t) + bx(t+1) + f(t), \tag{3.52}$$

where $b \in \mathbb{R}$ and f is 1-periodic and continuous. In this case, $\mathcal{B} = bS(1)$. Hence, $sp(f) = 2\pi\mathbf{Z}$ and condition (3.51) can be written as

$$1 \notin \sigma(e^A)e^b. \tag{3.53}$$

Hence, if condition (3.53) holds true, then Eq.(3.52) has a unique 1-periodic mild solution.

3.3 Decomposition Theorem and Periodic, Almost Periodic Solutions

We begin this section with several classical examples showing that in the resonant case when the conditions of the type $\sigma(A) \cap i\mathbb{R} \neq \emptyset$ are not satisfied, the problem of finding almost periodic solutions becomes much more complicated.

Example 3.4. Consider the simplest form of inhomogeneous equations when $A = 0$. In this case, actually we are concerned with conditions for the integral

$$F(t) := \int_0^t f(\xi)d\xi, \quad t \in \mathbb{R}$$

is τ-periodic, where f is assumed to be a scalar τ-periodic continuous function. A simple computation shows that F is of the form

$$F(t) = mt + G(t),$$

where

$$m := \frac{1}{\tau}\int_0^\tau f(\xi)d\xi$$

$$G \text{ is } \ \text{a } \tau\text{-periodic function.}$$

Thus F is τ-periodic if and only if $m = 0$, or equivalently, F is bounded on $\mathbb{R}$ (actually on any half line). That is, a boundedness condition is needed.

In the infinite dimensional case, the boundedness condition of the above type is not sufficient. Further conditions on the geometry of the Banach space, in which the equation is studied, is needed.

Example 3.5. Let c_0 be the Banach space of numerical sequences $\xi := \{\xi_n\}_{n\in\mathbb{N}} \subset \mathbb{C}$ such that $\lim_{n\to\infty}\xi_n = 0$ with $\|\xi\| := \sup_{n\in\mathbb{N}}|\xi_n|$. Let

$$f(t) := \{1/n\cos t/n\}_{n\in\mathbb{N}}, \quad F(t) = \int_0^t f(\eta)d\eta = \{\sin t/n\}_{n\in\mathbb{N}}.$$

One can shows that the range of the function F is not precompact, and so the function F cannot be almost periodic. In fact, suppose that the range of F is precompact. Then, consider the sequence of functionals $\phi_n \in c_0^*$ defined by the formula $\phi_n(\xi) = \xi_n$ if $\xi = \{\xi_n\}_{n\in\mathbb{N}}$. If the range $R(F)$ of F is precompact, then the convergence $\phi_n(\xi)$ would be uniform for $\xi \in R(F)$. But $\phi_n(F(t)) = \sin(t/n) \to 0$ non-uniformly with respect to $t \in \mathbb{R}$. This contradicts the assumption on the precompactness of $R(F)$.

To see how the resonance influences on the amplitude of oscillation we consider the following example.

Example 3.6. Consider the equation

$$\ddot{x}(t) + x(t) = \sin t, \quad t \in \mathbb{R},\ x(t) \in \mathbb{R}.$$

One can show that the function $x_1(t) = -\frac{t}{2}\cos t$ is a solution of this equation. Below is the graph of this solution on the interval $[-60, 60]$ which exhibits its unboundedness of the amplitude as time tends to infinity.

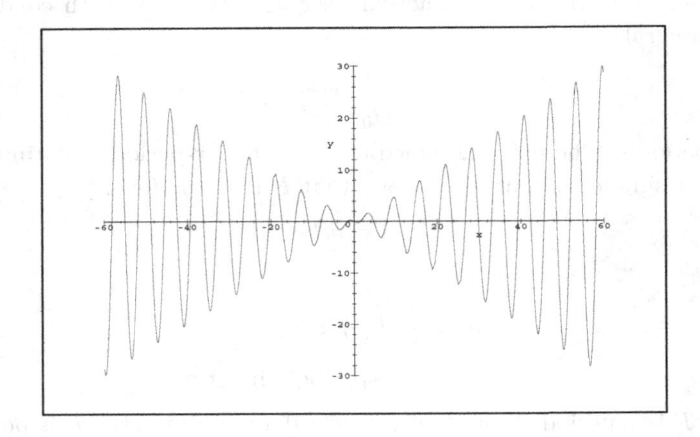

And any solution of the above equation is of the form

$$x(t) = C_1 \cos t + C_2 \sin t + x_1(t).$$

Hence any solution of the equation is unbounded, so it is not almost periodic.

We consider in this section the following linear inhomogeneous integral equation

$$x(t) = U(t,s)x(s) + \int_s^t U(t,\xi)g(\xi)d\xi, \forall t \geq s; t, s \in \mathbb{R}, \tag{3.54}$$

where f is continuous, $x(t) \in \mathbb{X}$, $\mathbb{X}$ is a Banach space, $(U(t,s))_{t \geq s}$ is assumed to be a 1-periodic evolutionary process on $\mathbb{X}$. As is known, continuous solutions of Eq.(3.54) correspond to the *mild solutions* of evolution equations

$$\frac{dx}{dt} = A(t)x + f(t), t \in \mathbb{R}, x \in \mathbb{X}, \tag{3.55}$$

where $A(t)$ is a (in general, unbounded) linear operator for every fixed t and is 1-periodic in t, and $(U(t,s))_{t\geq s}$ is generated by Eq.(3.55).

In the previous sections we have studied conditions for the existence and uniqueness (in some classes of function spaces) of almost periodic solutions of Eq.(3.54). In fact we have shown that if the following *nonresonant condition* holds

$$(\sigma(P) \cap S^1) \cap \overline{e^{isp(f)}} = \oslash, \tag{3.56}$$

where $P := U(1,0)$, S^1 denotes the unit circle of the complex plane, and f is almost periodic, then there exists an almost periodic solution x_f to Eq.(3.54) which is unique if one requires

$$\overline{e^{isp(x_f)}} \subset \overline{e^{isp(f)}}.$$

We may ask a question as what happens in *the resonant case* where condition (3.56) fails. Historically, this question goes back to a classical result of ordinary differential equations saying that supposing the finite dimension of the phase space $\mathbb{X}$ and the 1-periodicity of f Eq.(3.54) has a 1-periodic solution if and only if it has a bounded solution (see e.g. Theorem 20.3, p. 278 in [Amann (1)]). It is the purpose of this section to give an answer to the general problem as mentioned above (*Massera-type problem*): *Let Eq.(3.54) have a bounded (uniformly continuous) solution x_f with given almost periodic forcing term f. Then, when does Eq.(3.54) have an almost periodic solution w (which may be different from x_f) such that*

$$\overline{e^{isp(w)}} \subset \overline{e^{isp(f)}} \quad ?$$

In connection with this problem we note that various conditions are found on the bounded solution, itself, and the countability of the part of spectrum $\sigma(P) \cap S^1$ so that the bounded solution itself is almost periodic, or more generally, together with f belongs to a given function space $\mathcal{F}$. Here we note that this philosophy in general does not apply to the Massera-type problem for almost periodic solutions. In fact, it is not difficult to give a simple example in which f is 1-periodic and a bounded (uniformly continuous) solution to Eq.(3.54) exists, but this bounded solution itself is not 1-periodic.

Our method is to use the evolution semigroup associated with the process $(U(t,s))_{t\geq s}$ to study the harmonic analysis of bounded solutions to Eq.(3.54). As a result we will prove a spectral decomposition theorem for bounded solutions (Theorem 3.12 and Theorem 3.13) which seems to be useful in dealing with the above Massera-type problem. In fact, we will apply the spectral decomposition theorem to find new spectral criteria for the existence of almost periodic solutions and will consider particular cases to show the usefulness of this spectral decomposition technique. More concretely, even in the case where condition (3.56) fails we can still prove the existence of a bounded uniformly continuous solution w to Eq.(3.54) such that $\overline{e^{isp(w)}} = \overline{e^{isp(f)}}$ provided that $(\sigma(P) \cap S^1)\backslash\overline{e^{isp(f)}}$ is closed, and that Eq.(3.54) has a bounded uniformly continuous solution u (Corollary 3.8). Since w is a "spectral component" of u in case u is almost periodic the Fourier series of w is part of that of u (Corollary 3.9). Our Corollary 3.10 will deal with a particular autonomous case in which Corollary 3.8 fails to give a spectral criterion for the existence of quasi-periodic mild solutions.

For the sake of simplicity of notations we will use throughout the section the following notation: $\sigma(g) := \overline{e^{isp(g)}}$ for every bounded uniformly continuous function g. Throughout the section we will denote by $\sigma_\Gamma(P) = \sigma(P) \cap S^1$. Throughout this section $(U(t,s))_{t\geq s}$ will be assumed to be a 1-periodic strongly continuous evolutionary process. The operator $U(1,0)$ will be called the *monodromy operator* of the evolutionary process $(U(t,s))_{t\geq s}$ and will be denoted by P throughout this section. Note that the period of the evolutionary processes is assumed to be 1 merely for the sake of simplicity. Recall that

Definition 3.13. Let $(U(t,s))_{t\geq s}, (t,s \in \mathbb{R})$ be a 1-periodic strongly continuous evolutionary process and $\mathbf{F}$ be a closed subspace of $BUC(\mathbb{R},\mathbb{X})$ such that for every fixed $h > 0, g \in \mathbf{F}$ the map $t \mapsto U(t,t-h)g(t-h)$ belongs to $\mathbf{F}$. Then the semigroup of operators $(T^h)_{h\geq 0}$ on $\mathbf{F}$, defined by the formula

$$T^h g(t) = U(t,t-h)g(t-h), \forall t \in \mathbb{R}, h \geq 0, g \in \mathbf{F},$$

is called *evolution semigroup* associated with the process $(U(t,s))_{t\geq s}$ on $\mathbf{F}$.

We refer the reader to the previous section for further information on this semigroup. In the case where the evolution semigroup $(T^h)_{h\geq 0}$ is strongly continuous on $\mathbf{F}$ the explicit formula for the generator $\mathcal{A}$ of $(T^h)_{h\geq 0}$ is as follows:

Lemma 3.13. *Let $(T^h)_{h\geq 0}$ be strongly continuous on $\mathbf{F}$, a closed subspace of $BUC(\mathbb{R},\mathbb{X})$. Then its generator $\mathcal{A}$ is the operator on $\mathbf{F}$ with $D(\mathcal{A})$ consisting of all $g \in \mathbf{F}$ such that g is a solution to Eq.(3.54) with some $f \in \mathbf{F}$ (in this case such a function f is unique), by definition, $\mathcal{A}g = -f$.*

Proof. See the proof of Lemma 3.1. □

3.3.1 *Spectral Decomposition*

Let us consider the subspace $\mathcal{M} \subset BUC(\mathbb{R}, \mathbb{X})$ consisting of all functions $v \in BUC(\mathbb{R}, \mathbb{X})$ such that

$$\overline{e^{isp(v)}} := \sigma(v) \subset S_1 \cup S_2 \ , \tag{3.57}$$

where $S_1, S_2 \subset \mathbf{S}^1$ are disjoint closed subsets of the unit circle. We denote by $\mathcal{M}_v = \overline{spann\{S(t)v, t \in \mathbb{R}\}}$, where $(S(t))_{t\in\mathbb{R}}$ is the translation group on $BUC(\mathbb{R}, \mathbb{X})$, i.e. $S(t)v(s) = v(t+s), \forall t, s \in \mathbb{R}$.

Theorem 3.11. *Under the above notation and assumptions the function space $\mathcal{M}$ can be split into a direct sum $\mathcal{M} = \mathcal{M}_1 \oplus \mathcal{M}_2$ such that $v \in \mathcal{M}_i$ if and only if $\sigma(v) \subset S_i$ for $i = 1, 2$.*

Proof. Let $v \in \mathcal{M}$. Then, as is known (see Chapter 1)

$$isp(v) = \sigma(\mathcal{D}_{\mathcal{M}_v}). \tag{3.58}$$

Thus, by the Weak Spectral Mapping Theorem (see Chapter 1)

$$\sigma(S(1)|_{\mathcal{M}_v}) = \overline{e^{\sigma(\mathcal{D}_{\mathcal{M}_v})}} = \sigma(v) \subset S_1 \cup S_2. \tag{3.59}$$

Hence there is a spectral projection in $\mathcal{M}_v$ (note that in general we do not claim that this projection is defined on the whole space $\mathcal{M}$)

$$P_v^1 := \frac{1}{2i\pi} \int_\gamma R(\lambda, S(1)|_{\mathcal{M}_v}) d\lambda \ ,$$

where γ is a contour enclosing S_1 and disjoint from S_2, (or in general a union of fintely many such countours) by which we have

$$\sigma(P_v^1 S(1) P_v^1) \subset S_1. \tag{3.60}$$

On the other hand, denote $\Lambda_i \subset BUC(\mathbb{R}, \mathbb{X})$ consisting of all functions u such that $\sigma(u) \subset S_i$ for $i = 1, 2$. Then obviously, $\Lambda_i \subset \mathcal{M}$. Moreover, they are closed subspaces of $\mathcal{M}$, $\Lambda_1 \cap \Lambda_2 = \{0\}$. Now we show that if $v \in \mathcal{M}$, then $P_v^1 v \in \Lambda_1$ and $v - v_1 := v_2 \in \Lambda_2$. If this is true, then it yields that

$$\mathcal{M} = \Lambda_1 \oplus \Lambda_2.$$

To this end, we will prove

$$\sigma(v_j) \subset S_j, \forall j = 1, 2. \tag{3.61}$$

In fact we show that $\mathcal{M}_{v_1} = ImP_v^1$. Obviously, in view of the invariance of ImP_v^1 under translations we have $\mathcal{M}_{v_1} \subset ImP_v^1$. We now show the inverse. To this end, let $y \in ImP_v^1 \subset \mathcal{M}_v$. Then, by definition, we have

$$y = \lim_{n\to\infty} x_n,$$

where x_n can be represented in the form

$$x_n = \sum_{k=1}^{N(n)} \alpha_{k,n} S(t_{k,n}) v, \alpha_{k,n} \in \mathbf{C}, t_{k,n} \in \mathbb{R} \ \forall n.$$

Hence, since $y, x_n \in \mathcal{M}_v$

$$\begin{aligned} y = P_v^1 y &= \lim_{n\to\infty} \sum_{k=1}^{N(n)} \alpha_{k,n} S(t_{k,n}) P_v^1 v \\ &= \lim_{n\to\infty} \sum_{k=1}^{N(n)} \alpha_{k,n} S(t_{k,n}) v_1 \ . \end{aligned} \tag{3.62}$$

This shows that $y \in \mathcal{M}_{v_1}$. Thus, by the Weak Spectral Mapping Theorem and (3.60),

$$\overline{e^{isp(v_1)}} = \sigma(S(1)|_{\mathcal{M}_{v_1}}) = \sigma(S(1)|_{ImP_v^1}) \subset S_1.$$

By definition, $v_1 \in \Lambda_1$ and similarly, $v_2 \in \Lambda_2$. Thus the theorem is proved.

□

Remark 3.7. Below for every $v \in \mathcal{M}$ we will call $v_j, j = 1, 2$, as defined in the proof of Theorem 3.11, *spectral components* of the functions v. It is easily seen that if in the proof of Theorem 3.11, v is assumed to be almost periodic, then both spectral components v_j are almost periodic.

We will need the following lemma in the sequel

Lemma 3.14. *Let f be in $BUC(\mathbb{R}, \mathbb{X})$ and $(U(t,s))_{t\geq s}$ be a 1-periodic strongly continuous evolutionary process. Then the following assertions hold:*

(i) If $T : \mathbb{R} \to L(\mathbb{X})$ be 1-periodic and strongly continuous, then

$$\sigma(T(\cdot)f(\cdot)) \subset \sigma(f).$$

(ii) If f is of precompact range, then

$$\sigma\left(\int_t^{t+1} U(t+1,\xi) f(\xi) d\xi\right) \subset \sigma(f).$$

Proof. (i) Let T_n be the nth Cesaro mean of the Fourier series of T, so T_n is 1-periodic trignometric polynomial with value in $L(\mathbb{X})$ and $\|T_n(s)\| \leq \sup_{0\leq t\leq 1} \|T(t)\|$ and $T_n(s)x \to T(s)x$ uniformly in s for fixed $x \in \mathbb{X}$. For every n it is easily seen that $\sigma(T_n(\cdot)f(\cdot)) \subset \sigma(f)$. Set $\Lambda := \{\lambda \in \mathbb{R} : e^{i\lambda} \in$

$\sigma(f)\}$. Obviously, Λ is closed and $sp(T_n(\cdot)f(\cdot) \subset \Lambda$. Thus, if $\phi \in L^1(\mathbb{R})$ and $(supp\hat{\phi}) \cap \Lambda = \oslash$, then

$$0 = \int_{-\infty}^{\infty} \phi(t-s)T_n(s)f(s)ds \to \int_{-\infty}^{\infty} \phi(t-s)T_(s)f(s)ds$$

as $n \to \infty$, by Dominated Convergence Theorem. Thus

$$\int_{-\infty}^{\infty} \phi(t-s)T_(s)f(s)ds = 0$$

for all such ϕ. This proves (i).
(ii) Since f is of precompact range the evolution semigroup $(T^h)_{h\geq 0}$ associated with the process $(U(t,s))_{t\geq s}$ is strongly continuous at f. Thus, in view of (i)

$$\sigma\left(\int_0^h T^\xi f d\xi\right) \subset (f), \ \forall h \geq 0.$$

On the other hand

$$\int_0^1 T^\xi f d\xi(t+1) = \int_0^1 U(t+1, t+1-\xi)f(t+1-\xi)d\xi = \int_t^{t+1} U(t+1,\eta)f(\eta)d\eta.$$

This proves (ii). □

Lemma 3.15. *Let u be a bounded uniformly continuous solution to (3.54) and f be of precompact range. Then the following assertions hold true:*

(i)

$$\sigma(u) \subset \sigma_\Gamma(P) \cup \sigma(f), \tag{3.63}$$

(ii)

$$\sigma(u) \supset \sigma(f) \ . \tag{3.64}$$

Proof. (i) Set $P(t) := U(t, t-1)$, $\forall t \in \mathbb{R}$, $G := \{\lambda \in \mathbf{C} : e^\lambda \in \rho(P)\}$.

$$g(t) := \int_t^{t+1} U(t+1,\xi)f(\xi)d\xi, t \in \mathbb{R}.$$

By Lemma 3.14 $\sigma(g) \subset \sigma(f)$. By the definition of Carleman spectrum,

$$\hat{u}(\lambda) = \begin{cases} \int_0^\infty e^{-\lambda t}u(t)dt, (Re\lambda > 0); \\ \\ -\int_0^\infty e^{\lambda t}u(-t)dt, (Re\lambda < 0). \end{cases}$$

Hence, for $Re\lambda > 0$ and $\lambda \in G$ we have

$$\begin{aligned}
\hat{u}(\lambda) &= \int_0^\infty e^{-\lambda t}u(t)dt, \\
&= \int_0^\infty e^{-\lambda t}R(e^\lambda, P(t))(e^\lambda - P(t))u(t)dt \\
&= \int_0^1 e^{-\lambda t}e^\lambda R(e^\lambda, P(t))u(t)dt \\
&\quad + \int_0^\infty e^{-\lambda t}R(e^\lambda, P(t))(u(t+1) - P(t)u(t))dt \\
&= H(\lambda) + \int_0^\infty e^{-\lambda t}R(e^\lambda, P(t))g(t)dt, \qquad (3.65)
\end{aligned}$$

where

$$H(\lambda) := \int_0^1 e^{-\lambda t}e^\lambda R(e^\lambda, P(t))u(t)dt.$$

Obviously, $H(\lambda)$ is analytic in G. On the other hand, since $R(e^\lambda, P(t))$ is 1-periodic strongly continuous (see Lemma 3.3), by Lemma 3.14 the function $g_1(t) := R(e^\lambda, P(t))g(t)$ has the property that $\sigma(g_1) \subset \sigma(f)$. Thus from (3.65), for $Re\lambda > 0$, $\lambda \in G$,

$$\hat{u}(\lambda) = H(\lambda) + \hat{g_1}(\lambda). \qquad (3.66)$$

Finally if $\zeta_0 \in \mathbb{R} : e^{\zeta_0} \notin \sigma_\Gamma(P) \cup \sigma(f)$, then $\hat{u}$ has an analytic continuation at ζ_0. This completes the proof of (i).

(ii) Under the assumptions it may be seen that the evolution semigroup $(T^h)_{h\geq 0}$ associated with $(U(t,s))_{t\geq s}$ is strongly continuous at the function$u \in BUC(\mathbb{R}, \mathbb{X})$ (this can be checked directly using Eq.(3.54)), and f. Hence, by Lemma 3.13

$$\lim_{h\to 0^+} \frac{T^h u - u}{h} = \mathcal{A}u = -f \ . \qquad (3.67)$$

Hence, to prove (3.64) it suffices to show that $\sigma(T^h u) \subset \sigma(u)$. In turn, this is clear in view of Lemma 3.14. □

We are now in a position to state the main result of this section.

Theorem 3.12. *(Spectral Decomposition Theorem) Let u be a bounded, uniformly continuous solution to Eq.(3.54). Moreover, let f have pre-compact range and the sets $\sigma(f)$ and $\sigma(P) \cap S^1$ be contained in a disjoint union of the closed subsets $S_1, \cdots, S_k$ of the unit circle. Then the solution u can be decomposed into a sum of k spectral components*

$u_j, j = 1, \cdots, k$ such that each $u_j, j = 1, \cdots, k$ is a solution to Eq.(3.54) with $f = f_j, j = 1, \cdots, k$, respectively, where $f = \sum_{j=1}^k f_j$ is the decomposition of f into the sum of spectral components as described in Theorem 3.11, i.e. $u = \sum_{j=1}^k u_j, \sigma(u_j), \sigma(f_j) \subset S_j, j = 1, \cdots, k$ and $u_j \in BUC(\mathbb{R}, \mathbb{X})$ is a solution to Eq.(3.54) with $f := f_j$ for $j = 1, \cdots, k$.

Proof. Let us denote by $\mathcal{N}$ the subspace of $BUC(\mathbb{R}, \mathbb{X})$ consisting of all functions u such that $\sigma(u) \subset \cup_{j=1}^k S_j$. Then, by assumptions and Theorem 3.11 there are corresponding spectral projections $P_1, \cdots, P_k$ on $\mathcal{N}$ with properties that

(1) $P_j P_n = 0$ if $j \neq n$,
(2) $\Sigma_{j=1}^k P_j = I$,
(3) If $u \in Im P_j$, then $\sigma(P_j u) \subset S_j$ for all $j = 1, \cdots, k$.

Note that by Lemma 3.15 for every positive h and $j = 1, \cdots, k$ the operator T^h leaves $Im P_j$ invariant. Hence, $\mathcal{N}$ and $Im P_1, \cdots, Im P_k$ are invariant under the semigroup $(T^h)_{h \geq 0}$. Consequently, since u is a solution to Eq.(3.54) and f has precompact range the evolution semigroup $(T^h)_{h \geq 0}$ is strongly continuous at u and f. Using the explicit formula for the generator of $(T^h)_{h \geq 0}$ as described in Lemma 3.13 we have

$$\begin{aligned} P_j f = P_j \lim_{h \to 0^+} \frac{T^h u - u}{h} &= P_j \lim_{h \to 0^+} \Sigma_{n=1}^k P_n \frac{T^h u - u}{h} \\ &= \lim_{h \to 0^+} \frac{T^h P_j u - P_j u}{h}. \end{aligned} \tag{3.68}$$

This yields that $P_j u$ is a solution to Eq.(3.54) with corresponding $f_j = P_j f$.
□

Remark 3.8. If in Theorem 3.12 we assume furthermore that f and u are both almost periodic, then the spectral components $u_j, j = 1, \cdots, k$ are all almost periodic. This is not the case if neither u, nor f is almost periodic. However, if we have some additional information on the spectral sets S_j, e.g., their countability and the phase space $\mathbb{X}$ does not contain c_0, then the almost periodicity of u_j are guaranteed.

Now we are going to focus our special attention on autonomous equations of the form

$$dx/dt = Ax + f(t), \tag{3.69}$$

where A is the generator of a C_0-semigroup $(T(t))_{t\geq 0}$, and $f \in BUC(\mathbb{R}, \mathbb{X})$ has precompact range. Below we will use the following notation: $\sigma_i(A) = \{\lambda \in \mathbb{R} : i\lambda \in (\sigma(A) \cap i\mathbb{R})\}$. *By mild solutions* of Eq.(3.69) we will understand in a standard way that they are solutions to Eq.(3.54) with $U(t,s) := T(t-s), \forall t \geq s$. As shown below, in this case we can refine the spectral decomposition technique to get stronger assertions which usefulness will be shown in the next subsection when we deal with quasi-periodic solutions. To this purpose, we now prove the following lemma.

Lemma 3.16. *Let Eq.(3.69) satisfy the above conditions, i.e., A generates a C_0-semigroup and $f \in BUC(\mathbb{R}, \mathbb{X})$ has precompact range. Moreover, let u be a bounded uniformly continuous mild solution to Eq.(3.69). Then the following assertions hold:*

(i)

$$sp(u) \subset \sigma_i(A) \cup sp(f), \tag{3.70}$$

(ii)

$$sp(u) \supset sp(f). \tag{3.71}$$

Proof. (i) For (3.70) we compute the Carleman transform of u. For $Re\lambda > 0$,

$$\begin{aligned} \hat{u}(\lambda) &= \int_0^\infty e^{-\lambda t} u(t) dt, \\ &= \int_0^\infty e^{-\lambda t} T(t) u(0) dt + \int_0^\infty e^{-\lambda t} \left(\int_0^t T(t\xi) f(\xi) d\xi \right) dt \\ &= R(\lambda, A) u(0) + \int_0^\infty e^{-\lambda t} \left(\int_0^t T(t\xi) f(\xi) d\xi \right) dt. \end{aligned} \tag{3.72}$$

Note that in the same way as in the proof of Lemma 3.14 we can easily show that

$$sp\left(t \to \int_0^t T(t-\xi) f(\xi) d\xi\right) \subset sp(f). \tag{3.73}$$

Hence, (3.70) is proved.
(ii) Note that since for every $h > 0$ the operator T^h is a multiplication by a bounded operator $T(h)$ we have $sp(T^h u) \subset sp(u)$. Thus, using the argument of the proof of Lemma 3.15 (ii) we have

$$\begin{aligned} sp(f) = sp(-f) &= sp\left(\lim_{h\to 0+} \frac{T^h u - u}{h}\right) \\ &\subset sp(u) . \end{aligned}$$

This completes the proof of (ii). □

The main result for the autonomous case is the following:

Theorem 3.13. *Let A generate a C_0-semigroup and $f \in BUC(\mathbb{R}, \mathbb{X})$ have precompact range. Moreover, let u be a bounded uniformly continuous mild solution to Eq.(3.69). Then the following assertions hold true:*

(i) If

$$\overline{e^{i\sigma_i(A)}} \backslash \sigma(f)$$

is closed, Eq.(3.69) has a bounded uniformly continuous mild solution w such that $\sigma(w) = \sigma(f)$,

(ii) If $\sigma_i(A)$ is bounded and

$$\sigma_i(A) \backslash sp(f) \tag{3.74}$$

is closed, then Eq.(3.69) has a bounded uniformly continuous mild solution w such that $sp(w) = sp(f)$.

Proof. (i) Note that in this case together with (3.70) the proof of Theorem 3.12 applies.
(ii) Under the assumptions there exists a continuous function ψ which belongs to the Schwartz space of all C^∞-functions on $\mathbb{R}$ with each of its derivatives decaying faster than any polynomial such that its Fourier transform $\tilde{\psi}$ has $\sigma_i(A) \backslash sp(f)$ as its support (which is compact in view of the assumptions). Hence, every bounded uniformly continuous function g such that $sp(g) \subset \sigma_i(A) \cup sp(f)$ can be decomposed into the sum of two spectral components as follows:

$$g = g_1 + g_2 = \psi * g + (g - \psi * g),$$

where $g_1 = \psi * g, g_2 = (g - \psi * g)$. Moreover, this decomposition is continuous in the following sense: If $g^{(n)}, n = 1, 2, \cdots$ is a sequence in $BUC(\mathbb{R}, \mathbb{X})$ with $sp(g^{(n)}) \subset \sigma_i(A) \cup sp(f)$ such that $\lim_n g^{(n)} = g$ in $BUC(\mathbb{R}, \mathbb{X})$, then $\lim_n g_1^{(n)} = g_1, \lim_n g_2^{(n)} = g_2$. Hence we have in fact proved a version of Theorem 3.11 which allows us to employ the proof of Theorem 3.12 for this assertion (ii). □

Remark 3.9.

(1) In view of the failure of the Spectral Mapping Theorem for general C_0-semigroups the condition in the assertion (i) is a little more general than that formulated in terms of $\sigma(T(1))$.

(2) If we know beforehand that u is almost periodic, then in the statement of Theorem 3.13 we can claim that the spectral component w is almost periodic.

3.3.2 *Spectral Criteria For Almost Periodic Solutions*

This subsection will be devoted to some applications of the spectral decomposition theorem to prove the existence of almost periodic solutions with specific spectral properties. In particular, we will revisit the classical result by Massera on the existence of periodic solutions as well as its extensions. To this end, the following notion will play the key role.

Definition 3.14. Let $\sigma(f)$ and $\sigma_\Gamma(P)$ be defined as above. We say that the set $\sigma(f)$ and $\sigma_\Gamma(P)$ satisfy *the spectral separation condition* if the set $\sigma_\Gamma(P)\backslash\sigma(f)$ is closed.

Corollary 3.8. *Let f be almost periodic, $\sigma(f)$ and $\sigma_\Gamma(P)$ satisfy the spectral separation condition. Moreover, let $\sigma(f)$ be countable and $\mathbb{X}$ not contain any subspace which is isomorphic to c_0. Then if there exists a bounded uniformly continuous solution u to Eq.(3.54), there exists an almost periodic solution w to Eq.(3.54) such that $\sigma(w) = \sigma(f)$.*

Proof. We define in this case $S_1 := \sigma(f)$, $S_2 := \sigma_\Gamma(P)\backslash\sigma(f)$. Then, by Theorem 3.12 there exists a solution w to Eq.(3.54) such that $\sigma(w) \subset \sigma(f)$. Using the estimate (3.64) we have $\sigma(w) = \sigma(f)$. In particular, since $\sigma(w)$ is countable and $\mathbb{X}$ does not contain c_0, w is almost periodic. □

Remark 3.10.

(1) If $\sigma(f)$ is finite, then $sp(w)$ is discrete. Thus, the condition that $\mathbb{X}$ does not contain any subspace isomorphic to c_0 can be dropped.
(2) In the case where $\sigma_\Gamma(P)$ is countable it is known that with additional ergodic conditions on u the solution u has "similar spectral properties" as f. However, in many cases it is not expected that the solution u itself has similar spectral properties as f as in the Massera-type problem (see [Massera (68)], [Chow and Hale (20)], [Shin and Naito (96)], [Naito, Minh, R. Miyazaki and Shin (75)] e.g.).
(3) In the case where P is compact (or merely $\sigma_\Gamma(P)$ is finite) the spectral separation condition is always satisfied. Hence, we have a natural extension of a classical result for almost periodic solutions. In this case see also Corollary 3.9 below.
(4) We emphasize that the solution w in the statement of Corollary 3.8 is a "$\sigma(f)$-spectral component" of the bounded solution u. This will be helpful to find the Fourier coefficients of w as part of those of u.

(5) In view of estimate (3.64) w may be seen as a "minimal" solution in some sense.

Corollary 3.9. *Let all assumptions of Corollary 3.8 be satisfied. Moreover, let $\sigma_\Gamma(P)$ be countable. Then if there exists a bounded uniformly continuous solution u to Eq.(3.54), it is almost periodic. Moreover, the following part of the Fourier series of u*

$$\Sigma b_\lambda e^{i\lambda t} \quad , b_\lambda = \lim_{T\to\infty} \frac{1}{2T}\int_{-T}^{T} e^{-i\lambda\xi}u(\xi)d\xi, \tag{3.75}$$

where $e^{i\lambda} \in \sigma(f)$, is again the Fourier series of another almost periodic solution to Eq.(3.54).

Proof. The assertion that u is almost periodic is standard in view of (3.63) (see Chapter 1). It may be noted that in the case u is almost periodic, the spectral decomposition can be carried out in the function space $AP(\mathbb{X})$ instead of the larger space $BUC(\mathbb{R}, \mathbb{X})$. Hence, we can decompose the solution u into the sum of two almost periodic solutions with spectral properties described in Theorem 3.12. Using the definition of Fourier series of almost periodic functions we arrive at the next assertion of the corollary. □

The next corollary will show the advantage of Theorem 3.13 which allows us to take into account the structure of $sp(f)$ rather than that of $\sigma(f)$. To this end, we introduce the following terminology. A set of reals S is said to have an *integer and finite basis* if there is a finite subset $T \subset S$ such that any element $s \in S$ can be represented in the form $s = n_1b_1 + \cdots + n_mb_m$, where $n_j \in \mathbf{Z}, j = 1, \cdots, m$, $b_j \in T, j = 1, \cdots, m$. If f is quasi-periodic and the set of its Fourier-Bohr exponents is discrete (which coincides with $sp(f)$ in this case), then the spectrum $sp(f)$ has an integer and finite basis (see p.48 in [Levitan and Zhikov (58)]). Conversely, if f is almost periodic and $sp(f)$ has an integer and finite basis, then f is quasi-periodic. We refer the reader to pp. 42-48 in [Levitan and Zhikov (58)] more information on the relation between quasi-periodicity and spectrum, Fourier-Bohr exponents of almost periodic functions.

Corollary 3.10. *Let all assumptions of the second assertion of Theorem 3.13 be satisfied. Moreover, assume that $\mathbb{X}$ does not contain c_0. Then if $sp(f)$ has an integer and finite basis, Eq.(3.69) has a quasi-periodic mild solution w with $sp(w) = sp(f)$.*

Proof. Under the corollary's assumptions the spectrum $sp(w)$ of the solution w, as described in Theorem 3.13, is in particular countable. Hence w is almost periodic. Since $sp(w) = sp(f)$, $sp(w)$ has an integer and finite basis. Thus w is quasi-periodic. □

Below we will consider some particular cases

Example 3.7. *Periodic solutions.*

If $\sigma(f) = \{1\}$ we are actually concerned with the existence of periodic solutions. Hence, Corollary 3.8 extends the classical result to a large class of evolution equations which has 1 as an isolated point of $\sigma_\Gamma(P)$. Moreover, Corollary 3.9 provides a way to approximate the periodic solution. In particular, suppose that $\sigma_\Gamma(P)$ has finitely many elements, then we have the following:

Corollary 3.11. *Let $\sigma_\Gamma(P)$ have finitely many elements $\{\mu_1, \cdots, \mu_N\}$ and $u(\cdot)$ be a bounded uniformly continuous solution to Eq.(3.54). Then it is of the form*

$$u(t) = u_0(t) + \sum_{k=1}^{N} e^{i\lambda_k t} u_k(t), \tag{3.76}$$

where u_0 is a bounded uniformly continuous mild 1-periodic solution to the inhomogeneous equation (3.54), $u_k, k = 1, \cdots, N$, are 1-periodic solutions to Eq.(3.54) with $f = -i\lambda_k u_k$, respectively, $v(t) = \sum_{k=1}^{N} e^{i\lambda_k t} u_k(t)$ is a quasi periodic solution to the corresponding homogeneous equation of Eq.(3.54) and $\lambda_1, \cdots, \lambda_N$ are such that $0 < \lambda_1, \cdots, < \lambda_N < 2\pi$ and $e^{i\lambda_j} = \mu_j, j = 1, \cdots, N$.

Example 3.8. *Anti-periodic solutions.*

An anti-periodic (continuous) function f is defined to be a continuous one which satisfies $f(t+\omega) = -f(t), \forall t \in \mathbb{R}$ and here $\omega > 0$ is given. Thus, f is $2-\omega$-periodic. It is known that, the space of anti-periodic functions f with antiperiod ω, which is denoted by $\mathcal{AP}(\omega)$, is a subspace of $BUC(\mathbb{R}, \mathbb{X})$ with spectrum

$$sp(f) \subset \{\frac{2k+1\omega}{,} k \in \mathbf{Z}\}.$$

Without loss of generality we can assume that $\omega = 1$. Obviously, $\sigma(f) = \{-1\}, \forall f \in \mathcal{AP}(\omega)$. In this case the spectral separation condition is nothing but the condition that $\{-1\}$ is an isolated point of $\sigma_\Gamma(P)$.

Example 3.9.

Let u be a bounded uniformly continuous solution to Eq.(3.54) with f 2-periodic. Let us define

$$F(t) = \frac{f(t) - f(t+1)}{2}, G(t) = \frac{f(t) + f(t+1)}{2}, \forall t \in \mathbb{R}.$$

Then, it is seen that F is 1-anti-periodic and G is 1-periodic. Applying Theorem 3.12 we see that there exist two solutions to Eq.(3.54) as two components of u which are 1-antiperiodic and 1-periodic with forcing terms F, G, respectively. In particular, the sum of these solutions is a 2-periodic solution of Eq.(3.54) with forcing term f.

Example 3.10.

Let A be a sectorial operator in a Banach space $\mathbb{X}$ and the map $t \mapsto B(t) \in L(\mathbb{X}^\alpha, \mathbb{X})$ be Hölder continuous and 1-periodic. Then, as shown in Theorem 7.1.3 in [Henry (46)] the equation

$$\frac{dx}{dt} = (-A + B(t))x, \tag{3.77}$$

generates a 1-periodic strongly continuous evolutionary process $(U(t,s))_{t \geq s}$. If, furthermore, A has compact resolvent, then the monodromy operator P of the process is compact. Hence, for every almost periodic function f the sets $\sigma(f)$ and $\sigma_\Gamma(P)$ always satisfy spectral separation condition. In Section 1 we have shown that if $\sigma_\Gamma(P) \cap \sigma(f) = \oslash$, then there is a unique almost periodic solution x_f to the inhomogeneous equation

$$\frac{dx}{dt} = (-A + B(t))x + f(t) \tag{3.78}$$

with property that $\sigma(x_f) \subset \sigma(f)$. Now suppose that $\sigma_\Gamma(P) \cap \sigma(f) \neq \oslash$. By Corollary 3.8, if u is any bounded solution (the uniform continuity follows from the boundedness of such a solution to Eq.(3.78)), then there exists an almost periodic solution w such that $\sigma(w) = \sigma(f)$. We refer the reader to [Henry (46)] and [Pazy (90)] for examples from parabolic differential equations which can be included into the abstract equation (3.78).

Example 3.11.

Consider the heat equation in materials

$$\begin{cases} v_t(t,x) = \Delta v(t,x) + f(t,x), t \in \mathbb{R}, x \in \Omega \\ \\ v(t,x) = 0, t \in \mathbb{R}, x \in \partial\Omega, \end{cases} \tag{3.79}$$

where $\Omega \subset \mathbb{R}^n$ denotes a bounded domain with smooth boundary $\partial\Omega$. Let $\mathbb{X} = L^2(\Omega), A = \Delta$ with $D(A) = W^{2,2}(\Omega) \cap W_0^{1,2}(\Omega)$. Then A is self-adjoint and negative definite (see e.g. [Pazy (90)]). Hence $\sigma(A) \subset (-\infty, 0)$. In particular $\sigma_i(A) = \oslash$. Eq.(22) now becomes

$$\frac{dv}{dt} = Av + f. \tag{3.80}$$

We assume further that $f(t, x) = a(t)g(x)$ where a is a bounded uniformly continuous real function with $sp(a) = \mathbf{Z} \cup \pi\mathbf{Z}$, $g \in L^2(\Omega), g \neq 0$. It may be seen that $\sigma(f) = S^1$ and $sp(f)$ has an integer and finite basis. Hence, Theorem 3.12 does not give any information on the existence of a solution w with specific spectral properties. However, in this case Theorem 3.13 applies, namely, if Eq.(3.79) has a bounded solution, then it has a quasi periodic solution with the same spectrum as f.

3.4 Comments and Further Reading Guide

3.4.1 *Further Reading Guide*

The almost periodicity of a bounded solution to the evolution equation $u'(t) = Au(t) + f(t)$, $t \in \mathbb{R}$ is an interesting topic in the asymptotic behavior of evolution equations. The method of sums of commuting operators has been widely used in the study of the existence and regularity of solutions on finite intervals. The reader can find more in [Prato and Grisvard (23); Pruss (91)]. There is another approach to the admissibility theory via an operator equation $AX - XB = C$. The reader can find details of this method in [Vu and Schuler (103); Schweiker (94)]. Recently, the method of Fourier multipliers (see [Weis (104)]) has been used to study the existence and regularity of periodic solutions of inhomogeneous equations in [Arendt and Bu (8); Keyantuo and Lizama (51)], the stability and control in [Latushkin and Räbiger (57)]. There are many deep results discussing the conditions for a bounded and uniformly continuous mild solution u of this equation to be almost periodic. These conditions are stated in terms of the countability of imaginary spectrum of A and the phase space not containing c_0. In this direction we refer the reader to [Arendt, Batty, Hieber and Neubrander (7); Levitan and Zhikov (58)] and the references therein for more details. The decomposition method was first studied in [Naito, Minh, R. Miyazaki and Shin (75); Naito, Minh and Shin (76)]. Massera type criteria for the existence of almost periodic solutions with the same structure of spectrum as f are studied in detail for general classes of abstract functional differential equations in [Furumochi (35); Murakami, Naito and Minh (71); Minh and Minh (72)]. In [Minh and Minh (72)] the countability of spectrum condition is relaxed considerably.

3.4.2 *Comments*

The method of evolution semigroups was first used in [Naito and Minh (74)] to study the existence and uniqueness of almost periodic solutions. Subsequently, it was used in [Batty, Hutter and Räbiger (14); Murakami, Naito and Minh (71); Naito, Minh, R. Miyazaki and Shin (75); Naito, Minh and Shin (76)] and others. Section 1 is mainly taken from [Naito and Minh (74); Murakami, Naito and Minh (71)]. Section 2 is the main part of [Murakami, Naito and Minh (71)]. Section 3 with the decomposition method is taken from [Naito, Minh and Shin (76)]. The upper spectral estimate (3.63) was first found by Zhikov (see [Levitan and Zhikov (58)]) for autonomous equations. It was extended to periodic equations in [Batty, Hutter and Räbiger (14)]. The lower spectral estimate (3.64) was first found in [Naito, Minh and Shin (76)].

Chapter 4

Almost Automorphic Solutions

In this chapter we will study conditions under which an evolution equation has an almost automorphic (classical/mild) solution. The concept of almost automorphic functions was introduced by S. Bochner three decades ago. However, only recently the interest in almost automorphic solutions of evolution equations has been resurgent. We begin this chapter with some fundamental results on almost automorphic solutions of evolution equations in Section 4.1. In the next section we will present a new method using invariant subspaces. An existence result based on Fixed Point Theorems for semilinear equations will be given in Section 4.3. One difficulty we have to overcome in this section is due to the fact that almost automorphic functions are not necessarily uniformly continuous. We deal with this situation in Sections 4.4 and 4.5 by using the method of sums of commuting operators and the notion of uniform spectrum recently introduced in [Diagana, N'Guérékata and Minh (28)]. In Section 4.5, we are concerned with a second-order linear evolution equation.

4.1 The Inhomogeneous Linear Equation

In this section we will consider in a (real or complex) Banach space $\mathbb{X}$ the differential equation

$$\dot{x}(t) = Ax(t) + f(t), \quad t \in \mathbb{R} \tag{4.1}$$

where $f \in AA(\mathbb{X})$, the Banach space of all almost automorphic functions $\mathbb{R} \mapsto \mathbb{X}$ (see Section 1.3.7) and A is a (generally unbounded, unless otherwise stated) linear operator with domain $D(A) \subset \mathbb{X}$. Our concern is to look at conditions under which (classical or mild) bounded solutions of (4.1) are also in $AA(\mathbb{X})$.

This problem was initially raised and solved by Bohr and Neugebauer in the case where $f \in AP(\mathbb{X})$ and $\mathbb{X}$ is a finite dimensional space. The almost automorphic version of this result is well-known.

We begin our study with the following result

Theorem 4.1. *If $A = \lambda$ is a complex number and $f \in AA(\mathbb{C}^n)$, then the solution $x(t)$ of (4.1) is in $AA(\mathbb{C}^n)$.*

Proof. We first consider the case where $\lambda = i\theta$, where $\theta \in \mathbb{R}$. In this case,

$$x(t) = e^{i\theta t}x(0) + \int_0^t e^{i\theta(t-\xi)} f(\xi)d\xi, \quad \forall t \in \mathbb{R}.$$

Therefore, from the boundedness of $x(\cdot)$ it follows the boundedness of the function $t \mapsto \int_0^t e^{i\theta(t-\xi)} f(\xi)d\xi$. Consequently, the function $t \mapsto \int_0^t e^{i\theta(-\xi)} f(\xi)d\xi$ is bounded. Observe that the function $\xi \mapsto e^{i\theta(-\xi)} f(\xi)d\xi$ is almost automorphic, and so the integral function $t \mapsto \int_0^t e^{i\theta(-\xi)} f(\xi)d\xi$ is almost automorphic. Finally, we get the almost automorphy of the bounded solution $x(\cdot)$.

If $\Re e\ \lambda \neq 0$, then $x(t)$ is equal to $x_1(t)$ if $\Re e\ \lambda > 0$, and $x_2(t)$ if $\Re e\ \lambda < 0$, where

$$x_1(t) = -\int_t^\infty e^{\lambda(t-s)} f(s)ds \quad t \in \mathbb{R},$$

$$x_2(t) = \int_{-\infty}^t e^{\lambda(t-s)} f(s)ds \quad t \in \mathbb{R}$$

Let us prove that $x_1(t) \in AA(\mathbb{C}^n)$. The proof of $x_2(t) \in AA(\mathbb{C}^n)$ is similar.

Use the change of variable $\sigma = t - s$ to obtain

$$x_1(t) = -\int_{-\infty}^0 e^{\lambda\sigma} f(t-\sigma)d\sigma \quad t \in \mathbb{R}$$

Let now $(\sigma_n^{'})$ be an arbitrary sequence of real numbers. Since $f \in AA(\mathbb{C}^n)$, there exists a subsequence (σ_n) of $(\sigma_n^{'})$ such that

$$g(t) = \lim_{n\to\infty} f(t + \sigma_n)$$

is well defined for each $t \in \mathbb{R}$, and

$$f(t) = \lim_{n\to\infty} g(t - \sigma_n)$$

for each $t \in \mathbb{R}$.

Now fix $t \in \mathbb{R}$. Then we get

$$\lim_{n\to\infty} f(t - \sigma + \sigma_n) = g(t - \sigma)$$

for each given $\sigma \in \mathbb{R}$. Clearly

$$x_1(t + \sigma_n) = -\int_{-\infty}^{0} e^{\lambda\sigma} f(t - \sigma + \sigma_n)\, d\sigma$$

We observe also that

$$\|e^{\lambda\sigma} f(t - \sigma + \sigma_n)| \leq e^{(\Re e\ \lambda)\sigma} \sup_{t\in\mathbb{R}} \|f(t)\|$$

and

$$\int_{-\infty}^{0} e^{(\Re e\lambda)\sigma} \sup_{t\in\mathbb{R}} \|f(t)\|\, d\sigma = \frac{1}{\Re e\lambda} \sup_{t\in\mathbb{R}} \|f(t)\| \ < \infty$$

We know also the g is a bounded and measurable function. Using now the Lebesgue's dominated convergence theorem, we obtain

$$\lim_{n\to\infty} x_1(t - \sigma_n) = -\int_{-\infty}^{0} e^{\lambda\sigma} g(t - \sigma)\, d\sigma = y_1(t)$$

for each $t \in \mathbb{R}$. We can apply the same reasoning to $y_1(t)$ to obtain

$$\lim_{n\to\infty} y_1(t - \sigma_n) = x_1(t)$$

for each $t \in \mathbb{R}$, which proves that $x_1(t) \in AA(\mathbb{C}^n)$ and completes the proof.

□

Now we have:

Theorem 4.2. *([N'Guérékata (79)][Remark 4.2.2]) If A is a linear operator $\mathbb{R}^n \mapsto \mathbb{R}^n$, then every bounded solution of (4.1) is in $AA(\mathbb{R}^n)$.*

Proof. First, we note that by Floquet Theory of periodic ordinary differential equations, without loss of generality we may assume that A is independent of t.

Next we will show that the problem can be reduced to the one-dimensional case. In fact, if A is independent of t, by a change of variable if necessary, we may assume that A is of Jordan normal form. In this direction we can go further with assumption that A has only one Jordan box. That is, we have to prove the theorem for equations of the form

$$\begin{pmatrix} \dot{x}_1(t) \\ \dot{x}_2(t) \\ \vdots \\ \dot{x}_n(t) \end{pmatrix} = \begin{pmatrix} \lambda & 1 & 0 & \dots & 0 \\ 0 & \lambda & 1 & \dots & 0 \\ \dots & \dots & \dots & \dots & \dots \\ 0 & 0 & 0 & \dots & \lambda \end{pmatrix} + \begin{pmatrix} f_1(t) \\ f_2(t) \\ \vdots \\ f_n(t) \end{pmatrix}.$$

Let us consider the last equation involving $x_n(t)$. We have

$$\dot{x}_n(t) = \lambda x_n(t) + f_n(t), \quad t \in \mathbb{R}, x(t) \in \mathbb{C}. \tag{4.2}$$

If $\Re\lambda \neq 0$, then we can easily check that either

$$y(t) = \int_{-\infty}^{t} e^{\lambda(t-\xi)} f_n(\xi)d\xi, \quad (\Re\lambda < 0)$$

or

$$z(t) = \int_{t}^{\infty} e^{\lambda(t-\xi)} f_n(\xi)d\xi, \quad (\Re\lambda > 0)$$

is a unique bounded solution of Eq. (4.2) Moreover, by Theorem 4.1, in both cases, $y(t)$ and $z(t)$ are in $AA(\mathbb{C})$. Hence, x_n is in $AA(\mathbb{C})$.

If $\Re\lambda = 0$, then $\lambda = i\eta$ for $\eta \in \mathbb{R}$. By assumption, there is a constant c such that the function

$$x_n(t) := ce^{i\eta t} + \int_0^t e^{i\eta(t-\xi)} f(\xi)d\xi,$$

is bounded on $\mathbb{R}$. This yields the boundedness of $\int_0^t e^{-i\eta\xi} f(\xi)d\xi$ on $\mathbb{R}$. Hence, $\int_0^t e^{-i\eta\xi} f(\xi)d\xi$ is in $AA(\mathbb{X})$. Finally, this yields that x_n is in $AA(\mathbb{X})$.

Let us consider next the equation involving x_{n-1} and x_n. Since x_n is in $AA(\mathbb{X})$, by repeating the above argument we can show that x_{n-1} is also in $AA(\mathbb{X})$. Continuing this process, we can show that all $x_k(\cdot)$ are in $AA(\mathbb{X})$. The proof is completed. □

Suppose now $f \in C(\mathbb{R}, \mathbb{X}) \cap L^1(\mathbb{R}^+, \mathbb{X})$ and A is the infinitesimal generator of a C_0-group $(T(t))_{t\in\mathbb{R}}$.

Consider the function $x \in C(\mathbb{R}, \mathbb{X})$ defined by

$$x(t) = T(t)x(0) + \int_0^t T(t-s)f(s)ds, \quad t \in \mathbb{R}$$

Let us assume that $T(t)y \in AA(\mathbb{X})$ for every $y \in \mathbb{X}$, which implies that $T(-t)y \in AA(\mathbb{X})$ for every $y \in \mathbb{X}$.

Now we write

$$x(t) = u(t) + v(t), \quad t \in \mathbb{R}$$

where

$$u(t) = T(t)x(0) + \int_0^\infty T(t-s)f(s)ds$$

and

$$v(t) = -\int_t^\infty T(t-s)f(s)ds.$$

It is clear that $u \in AA(\mathbb{X})$ by Theorem 1.19 i) and $v \in C(\mathbb{R}^+, \mathbb{X})$ with $\lim_{t\to\infty} \|v(t)\| = 0$.

Finally $x(t)_{|_{t\in\mathbb{R}^+}}$ is an asymptotically almost automorphic function.

We continue our study of (4.1) with the following result ([N'Guérékata (82)]).

Theorem 4.3. *Let A be the infinitesimal generator of a C_0-group of bounded linear operators $(T(t))_{t\in\mathbb{R}}$. Assume that there exists a subspace $\mathbb{X}_1$ of $D(A)$ with* $\dim \mathbb{X}_1 < \infty$ *such that:*

i) $A\mathbb{X}_1 \subset \mathbb{X}_1$

ii) $(T(t) - I)f(s) \in \mathbb{X}_1, \quad \forall s, t \in \mathbb{R}$

iii) A commutes with the projection $P : \mathbb{X} \mapsto \mathbb{X}_1$, on $D(A)$.

Then every solution $x(t)$ of (4.1) with $x(0) \in \mathbb{X}_1$ and precompact range $R_x = \{x(t)/t \in \mathbb{R}\}$ is in $AA(\mathbb{X})$.

Proof. Basically the proof uses the so-called *method of decomposition of the space* in order to apply a result of the Bohr-Neugebauer type (Theorem 4.1).

We let $P : \mathbb{X} \mapsto \mathbb{X}_1$, be the natural projection on $\mathbb{X}_1$ and consider the associated decomposition

$$\mathbb{X} = \mathbb{X}_1 \oplus kerP$$

where $kerP = \{x \in \mathbb{X}/Px = 0\}$ is the kernel of P.

Let $Q = I - P$; then $Q\mathbb{X}_1 = \{0\}$ and $Q^2 = I$. We recall also that both P and Q are bounded linear operators.

Now if $x(t)$ is a solution of (4.1) as given in the Theorem, then we can write

$$x(t) = x_1(t) + x_2(t), \quad t \in \mathbb{R}$$

where $x_1(t) = Px(t) \in \mathbb{X}_1$ and $x_2(t) = Qx(t) \in kerP$ for each $t \in \mathbb{R}$. Since R_x is precompact, it is also clear that both R_{x_1} and R_{x_2} are precompact.

We have

$$x(t) = T(t)x(0) + \int_0^t T(t-s)f(s)ds =$$

$$T(t)x(0) + \int_0^t f(s)ds + \int_0^t (T(t-s) - I)f(s)ds$$

Note that $\int_0^t (T(t-s) - I)f(s)ds \in \mathbb{X}_1$, by ii) in the Theorem, hence $Q\int_0^t (T(t-s) - I)f(s)ds = 0$.

Consequently

$$x_2(t) = Qx(t) = QT(t)x(0) + Q\int_0^t f(s)ds = QT(t)x(0) + \int_0^t Qf(s)ds$$

Now

$$\dot{x}_2(t) = QAT(t)x(0) + Qf(t) = QT(t)Ax(0) + Qf(t) = Qf(t)$$

Since $Ax(0) \in \mathbb{X}_1$, that yields $T(t)Ax(0) \in \mathbb{X}_1$ and consequently $QT(t)Ax(0) = 0$.

By *Exercise 4.1* below, $Qf(t)$ is almost automorphic; so is $\dot{x}_2(t)$. Now since R_{x_2} is precompact, we deduce that x_2 is in $AA(\mathbb{X})$.

Now it suffices to prove that $x_1(t)$ is almost automorphic in order to reach the conclusion based on Theorem 1.19 i).

We rewrite (4.1) as follows

$$\dot{x}(t) = Ax_1(t) + Ax_2(t) + Pf(t) + Qf(t)$$

and we apply P to both sides of the previous equation to get

$$\dot{x}_1(t) = PAx_1(t) + PAx_2(t) + P^2f(t) + PQf(t)$$

$$= PAx_1(t) + APx_2(t) + P^2f(t) + PQf(t) = PAx_1(t) + g(t)$$

where $g(t) = P^2f(t) + PQf(t)$ is almost automorphic.
This last equation holds true in the finite dimensional space $\mathbb{X}_1$. It is of the Borh-Neugebauer type (*Theorem 4.1*). So its solution $x_1(t)$ is almost automorphic. The proof is achieved. □

Exercise 18. Let $f \in AA(\mathbb{X})$ and $A \in B(\mathbb{X})$, the space of all bounded linear operators on $\mathbb{X}$. Show that $Af \in AA(\mathbb{X})$.

Exercise 19. Let $\mathbb{H}$ be a Hilbert space and $A : \mathbb{H} \mapsto \mathbb{H}$ a linear operator such that $A\phi_n = \lambda_n\phi_n$ $n = 1, 2, 3, ...$ where $\{\phi_1, \phi_2, ...\}$ is an orthonormal base for $\mathbb{H}$ and $|\Re e\lambda_n| \leq M$ for all $n = 1, 2, 3, ...$ Find the subspace $\mathbb{X}_1$ of $\mathbb{H}$ such that $A\mathbb{X}_1 \subset \mathbb{X}_1$, and A commutes with the projection $P : \mathbb{H} \mapsto \mathbb{X}_1$.

4.2 Method of Invariant Subspaces and Almost Automorphic Solutions of Second-Order Differential Equations

In this section we are concerned with the almost automorphic solutions to the homogeneous second-order hyperbolic differential equation of the form

$$\frac{d^2}{ds^2}u(s) + 2B\,\frac{d}{ds}u(s) + A\,u(s) = 0, \tag{4.3}$$

and the associated nonhomogeneous differential equation

$$\frac{d^2}{ds^2}u(s) + 2B\,\frac{d}{ds}u(s) + A\,u(s) = f(s), \tag{4.4}$$

where A, B are densely defined closed linear operators acting in a Hilbert space $\mathbb{H}$ and $f \in AA(\mathbb{H})$.

We use *invariant subspaces theory* (see Appendix 6.4) to show that under appropriate assumptions, every solution to the equations (4.2) and/or (4.3) is an almost automorphic vector-valued function. The idea of using the method of invariant subspaces to study the existence of almost automorphic solutions is recent and due to Diagana and N'Guérékata.

Let us indicate that the invariant subspaces method works smoothly in the framework of abstract differential equations involving the algebraic sum of unbounded linear operators.

Now setting $v(s) = \frac{d}{ds}u(s)$, the problem (4.2)-(4.3) can be rewritten in $\mathbb{H} \times \mathbb{H}$ of the form

$$\frac{d}{ds}\mathcal{U}(s) = (\mathcal{A} + \mathcal{B})\,\mathcal{U}(s), \tag{4.5}$$

and

$$\frac{d}{ds}\mathcal{U}(s) = (\mathcal{A} + \mathcal{B})\,\mathcal{U}(s) + F(s), \tag{4.6}$$

where $\mathcal{U}(s) = (u(s), v(s))$, $F(s) = (0, f(s))$ and $\mathcal{A}, \mathcal{B}$ are the operator matrices of the form

$$\mathcal{A} = \begin{pmatrix} O & I \\ -A & O \end{pmatrix} \quad \text{and} \quad \mathcal{B} = \begin{pmatrix} O & O \\ O & -2B \end{pmatrix},$$

on $\mathbb{H} \times \mathbb{H}$ with $D(\mathcal{A}) = D(A) \times \mathbb{H}$, $D(\mathcal{B}) = \mathbb{H} \times D(B)$, and O, I denote the zero and identity operators on $\mathbb{H}$, respectively.

Since (4.2)-(4.3) is equivalent to (4.4)-(4.5), instead of studying (4.2)-(4.3), we will focus on the characterization of almost automorphic solutions to (4.4)-(4.5).

Throughout the section, $\mathbb{H}$, $D(C)$, $R(C)$ and $N(C)$, denote a Hilbert space, the domain, the range and the kernel of the linear operator C, respectively.

Let A and B be densely defined closed unbounded linear operators on $\mathbb{H}$. Recall that their algebraic sum is defined by

$$D(A+B) = D(A) \cap D(B) \text{ and } (A+B)x = Ax + Bx,$$

$\forall x \in D(A) \cap D(B)$.

Since both A and B are densely defined, then the algebraic sum of $\mathcal{A}$ and $\mathcal{B}$, $\mathcal{S} = \mathcal{A} + \mathcal{B}$ is also a densely defined operator and

$$D(\mathcal{A} + \mathcal{B}) = D(A) \times D(B), \text{ and } (\mathcal{A} + \mathcal{B})\mathcal{U} = \mathcal{A}\mathcal{U} + \mathcal{B}\mathcal{U}.$$

Throughout the section, $\mathcal{A}$ and $\mathcal{B}$ will play similar roles.

Setting our main result, we make the following assumptions:

The operators $\mathcal{A}$ and $\mathcal{B}$ are infinitesimal generators of C_0-groups of bounded operators $(\mathcal{T}(t))_{s\in\mathbb{R}}$, $(\mathcal{R}(t))_{s\in\mathbb{R}}$, respectively, such that

(i) $\mathcal{T}(s)\mathcal{U} : s \mapsto \mathcal{T}(s)\mathcal{U}$ is almost automorphic for each $\mathcal{U} \in \mathbb{H} \times \mathbb{H}$, $\mathcal{R}(s)\mathcal{V} : s \mapsto \mathcal{R}(s)\mathcal{V}$ is almost automorphic for each $\mathcal{V} \in \mathbb{H} \times \mathbb{H}$, respectively.

(ii) there exists $S \subset \mathbb{H} \times \mathbb{H}$, a closed subspace that reduces both $\mathcal{A}$ and $\mathcal{B}$.

We denote by P_S and $Q_S = (I \times I - P_S) = P_{[\mathbb{H}\times\mathbb{H}]\ominus S}$, the orthogonal projections onto S and $[\mathbb{H} \times \mathbb{H}] \ominus S$, respectively.

(iii) $R(\mathcal{A}) \subset R(P_S) = N(Q_S)$

(iv) $R(\mathcal{B}) \subset R(Q_S) = N(P_S)$

Remark 4.1.

1. Recall that if A, B generate C_0-groups, then their sum $\mathcal{A} + \mathcal{B}$ need not be a generator of a C_0-group .

2. The assumption (ii) implies that both S and $[\mathbb{H} \times \mathbb{H} \ominus S]$ are invariant for the algebraic sum $\mathcal{A} + \mathcal{B}$ (note that it is well-defined as stated above).

Now we state and prove:

Theorem 4.4. *Under assumptions (i)-(ii)-(iii)-(iv), every solution to the differential equation (4.4) is almost automorphic.*

Proof. Let $X(s)$ be a solution to (4.4). Clearly $X(s) \in D(\mathcal{A}) \cap \overline{D(\mathcal{B})} = D(A) \times D(B)$ (notice that the algebraic sum $\mathcal{S} = \mathcal{A} + \mathcal{B}$ exists since $\overline{D(S)} = \mathbb{H} \times \mathbb{H}$).

Now decompose $X(s)$ as follows

$$X(s) = P_S X(s) + (I \times I - P_S)X(s),$$

where $P_S X(s) \in R(P_S) = N(Q_S)$, and $Q_S X(s) \in N(P_S) = R(Q_S)$.

We have

$$\begin{aligned}\frac{d}{ds}(P_S X(s)) &= P_S \frac{d}{ds} X(s)\\ &= P_S \mathcal{A} X(s) + P_S \mathcal{B} X(s)\\ &= \mathcal{A} P_S X(s) + P_S \mathcal{B} X(s) \quad \text{(according to}(ii))\\ &= \mathcal{A} P_S X(s) \quad \text{(according to}(iv))\end{aligned}$$

From $\frac{d}{ds}(P_S X(s)) = \mathcal{A} P_S X(s)$, it follows that

$$P_S X(s) = \mathcal{T}(t) P_S X(0).$$

Now according to (i), the vector-valued function $s \mapsto P_S X(s) = \mathcal{T}(t) P_S X(0)$ is almost automorphic.

In the same way, since $[\mathbb{H} \times \mathbb{H}] \ominus ([\mathbb{H} \times \mathbb{H}] \ominus S) = S$. It follows that the closed subspace S reduces $\mathcal{A}$ and $\mathcal{B}$ if and only if $[\mathbb{H} \times \mathbb{H}] \ominus S$ does. In other words, $[\mathbb{H} \times \mathbb{H}] \ominus S$ reduces $\mathcal{A}$ and $\mathcal{B}$. That is, a similar remark as *Remark 6.1* in *Appendix* holds when S is replaced by $[\mathbb{H} \times \mathbb{H}] \ominus S$. Thus, we have

$$\begin{aligned}\frac{d}{ds}(Q_S X(s)) &= Q_S \frac{d}{ds} X(s)\\ &= Q_S \mathcal{A} X(s) + Q_S \mathcal{B} X(s)\\ &= Q_S \mathcal{A} X(s) + \mathcal{B} Q_S X(s) \quad \text{(according to}(ii))\\ &= \mathcal{B} Q_S X(s) \quad \text{(according to}(iii))\end{aligned}$$

From the equation $\frac{d}{dt}(Q_S X(s)) = \mathcal{B} Q_S X(s)$, it follows that $s \mapsto Q_S X(s) = \mathcal{R}(s) Q_S X(0)$ is almost automorphic (according to (i)).

Therefore $X(s) = P_S X(s) + Q_S X(s)$ is also almost automorphic as the sum of almost automorphic vector-valued functions. □

Corollary 4.1. *Let $B : \mathbb{H} \mapsto \mathbb{H}$ be a bounded linear operator in the Hilbert space $\mathbb{H}$. Assume that assumptions (i)-(ii)-(iii)-(iv) hold true.*

Then every solution to the equation (4.5) is almost automorphic.

Proof. This an immediate consequence of Theorem 4.5 to the case where $\mathcal{B}$ is a bounded linear operator, it is straightforward. □

Consider now the nonhomogeneous equation (4.5). Assume that the vector valued function $f: \mathbb{R} \mapsto \mathbb{H}$ is almost automorphic. In fact, this implies that $F: s \mapsto (0, f(s))$ is in $\mathbb{A}A(\mathbb{H} \times \mathbb{H})$.

We have

Theorem 4.5. *Under assumption (i)-(ii)-(iii)-(iv), assume in addition that $f \in \mathbb{A}A(\mathbb{H})$ and $\int_0^t \mathcal{T}(-s)f(s)ds$ and $\int_0^t \mathcal{R}(-s)f(s)ds$ are bounded over $\mathbb{R}$.*

Then every solution to the equation (4.5) is almost automorphic.

Proof. Let $X(s)$ be a solution to (4.5). As in the proof of *Theorem 4.5*, the solution $X(s) \in D(\mathcal{A}) \cap D(\mathcal{B})$. Now express $X(s)$ as $X(s) = P_S X(s) + Q_S X(s)$, where P_S, $Q_S = (I \times I - P_S) = P_{[\mathbb{H}\times\mathbb{H}]\ominus S}$ are the orthogonal projections defined above.

We have

$$\begin{aligned}\frac{d}{ds}(P_S X(s)) &= P_S \frac{d}{ds} X(s)\\ &= P_S \mathcal{A} X(s) + P_S \mathcal{B} X(s) + P_S F(s)\\ &= \mathcal{A} P_S X(s) + P_S \mathcal{B} X(s) + P_S F(s) \quad \text{(according to}(ii))\\ &= \mathcal{A} P_S X(s) + P_S F(s) \quad \text{(according to}(iv))\end{aligned}$$

From $\frac{d}{ds}(P_S X(s)) = \mathcal{A} P_S X(s) + P_S F(s)$; it follows that

$$P_S X(s) = \mathcal{T}(s) P_S X(0) + \int_0^s \mathcal{T}(s-\sigma) P_S F(\sigma) d\sigma.$$

Set $G(s) = \int_0^s \mathcal{T}(s-\sigma) P_S F(\sigma) d\sigma$.

First observe that $\sigma \mapsto \mathcal{T}(-\sigma)\ P_S\ F(\sigma)$ is almost automorphic. Moreover the function $x(s) =: \int_0^s \mathcal{T}(-\sigma) P_S F(\sigma) d\sigma$ is bounded (as it can be easily proved), thus it is almost automorphic. Now $\mathcal{T}(s)\ x(s) = G(s)$ is almost automorphic.

According to assumption (i), the vector-valued function

$$s \mapsto P_S X(s) = \mathcal{T}(s) P_S X(0)$$

is almost automorphic. Therefore $s \mapsto P_S X(s)$ is almost automorphic as the sum of almost automorphic vector-valued functions.

In the same way, it is not hard to see that

$$\frac{d}{ds}(Q_S X(s)) = \mathcal{B} Q_S X(s) + Q_S F(s),$$

and that $Q_S X(s)$ can be expressed as

$$Q_S X(s) = \mathcal{R}(s) Q_S X(0) + \int_0^s \mathcal{R}(s - \sigma) Q_S F(\sigma) d\sigma.$$

Using similar arguments as above, it can be shown that $s \mapsto Q_S X(s)$ is almost automorphic. Therefore $X(s) = P_S X(s) + Q_S X(s)$ is also an almost automorphic vector-valued function. □

Remark 4.2. Let us notice that the previous results (Theorem 4.5 and Theorem 4.6) still hold in the case where $\mathcal{A}, \mathcal{B} : \mathbb{H} \times \mathbb{H} \mapsto \mathbb{H} \times \mathbb{H}$ are bounded linear operator matrices on $\mathbb{H} \times \mathbb{H}$. In such a case, the similar assumptions are required, that is, (i)-(ii)-(iii) and (iv).

4.3 Existence of Almost Automorphic Solutions to Semilinear Differential Equations

This section is concerned with the differential equation in a Banach space $\mathbb{X}$:

$$x'(t) = Ax(t) + f(t, x(t)), \quad t \in \mathbb{R} \tag{4.7}$$

where A is the infinitesimal generator of an exponentially stable C_0-semigroup $(T(t))_{t \geq 0}$; that is, there exist $K > 0, \ \omega < 0$ such that

$$\|T(t)\| \leq K e^{\omega t}, \quad \forall t \geq 0.$$

We assume that $f : \mathbb{R} \times \mathbb{X} \mapsto \mathbb{X}$ satisfies a Lipschitz condition in x uniformly in t, that is,

$$\|f(t, x) - f(t, y)\| \leq L \|x - y\|$$

$\forall t \in \mathbb{R}, \ x, y \in \mathbb{X}$.

We like to establish existence and uniqueness of almost automorphic mild solutions to the Eq. (4.6). We first prove the existence of almost automorphic mild solution of the differential equation

$$x'(t) = Ax(t) + f(t), \quad t \in \mathbb{R}, \tag{4.8}$$

where f is almost automorphic.

Theorem 4.6. *Let $f \in AA(\mathbb{X})$. Then equation (4.7) has a unique almost automorphic mild solution.*

Proof. We first prove existence of an almost automorhic solution. It is well-known that Eq. (4.7) possesses a mild solution of the form

$$x(t) = T(t-a)x(a) + \int_a^t T(t-s)f(s)ds, \ for\ all\ a \in \mathbb{R},\ t \geq a.$$

It remains to prove that it is almost automorphic.

First, we consider the function $u(t) := \int_{-\infty}^t T(t-s)f(s)ds$, defined as

$$\int_{-\infty}^t T(t-s)f(s)ds = \lim_{r \to -\infty} \int_r^t T(t-s)f(s)ds.$$

Clearly for each $r < t$, the integral $\int_r^t T(t-s)f(s)ds$ exists. Moreover

$$\| \int_r^t T(t-s)f(s)ds\| \leq \frac{K}{|\omega|}\|f\|_\infty, \ \forall r < t.$$

which shows $\int_{-\infty}^t T(t-s)f(s)ds$ is absolutely convergent.

Now let (s_n') be an arbitrary sequence of real numbers. Since $f \in AA(\mathbb{X})$, there exists a subsequence (s_n) of (s_n') such that

$$g(t) = \lim_{n \to +\infty} f(t+s_n)$$

is well-defined for each $t \in \mathbb{R}$ and

$$f(t) = \lim_{n \to +\infty} g(t-s_n)$$

for each $t \in \mathbb{R}$

Now consider

$$\begin{aligned} u(t+s_n) &= \int_{-\infty}^{t+s_n} T(t+s_n-s)f(s)ds \\ &= \int_{-\infty}^t T(t-\sigma)f(\sigma+s_n)d\sigma \\ &= \int_{-\infty}^t T(t-\sigma)f_n(\sigma)d\sigma, \end{aligned}$$

where $f_n(\sigma) = f(\sigma+s_n), \ n = 1, 2, ...$

We also have

$$\|u(t+s_n)\| \leq \frac{K}{|\omega|}\|f\|_\infty, \ \forall n = 1, 2, ...$$

and by continuity of the semigroup, $T(t-\sigma)f_n(\sigma) \mapsto T(t-\sigma)g(\sigma)$, as $n \to \infty$ for each $\sigma \in \mathbb{R}$ fixed and any $t \geq \sigma$.

If we let $v(t) = \int_{-\infty}^{t} T(t-s)g(s)ds$, we observe that the integral is absolutely convergent for each t. So, by the Lebesgue's dominated convergent theorem,

$$u(t+s_n) \mapsto v(t), \quad \text{as} \quad n \to \infty$$

for each $t \in \mathbb{R}$.

We can show in a similar way that

$$v(t-s_n) \mapsto u(t) \quad \text{as} \quad n \to \infty$$

for each $t \in \mathbb{R}$. This shows that $u \in AA(\mathbb{X})$.

Now let $u(a) = \int_{-\infty}^{a} T(a-s)f(s)ds$. So $T(t-a)u(a) = \int_{-\infty}^{a} T(t-s)f(s)ds$.

If $t \geq a$, then

$$\begin{aligned}\int_a^t T(t-s)f(s)ds &= \int_{-\infty}^{t} T(t-s)f(s)ds - \int_{-\infty}^{a} T(t-s)f(s)ds \\ &= u(t) - T(t-a)u(a).\end{aligned}$$

so that, $u(t) = T(t-a)u(a) + \int_a^t T(t-s)f(s)ds$. If we fix $x(a) = u(a)$, then $x(t) = u(t)$, that is $x \in AA(\mathbb{X})$.

We finally prove the uniqueness of the almost aumorphic solution.

Assume x and y are two such solutions and we let $z = x - y$. Then $z \in AA(\mathbb{X})$ and satisfies the equation

$$z'(t) = Az(t), \quad t \in \mathbb{R}.$$

Note that z is bounded and satisfies also the equation

$$z(t) = T(t-s)z(s) \ \ \forall t, s \in \mathbb{R}, \quad t \geq s$$

We also have the inequality

$$\|z(t)\| \leq Ke^{\omega(t-s)}.$$

Take a sequence of real numbers (s_n) such that $s_n \to -\infty$. For any fixed $t \in \mathbb{R}$, we then can find a subsequence (s_{n_k}) of (s_n) with $s_{n_k} < t$ for all $k = 1, 2, ...$ Using the fact that $\omega < 0$, we obtain $z = 0$.

This shows uniqueness of the solution and ends the proof. □

Theorem 4.7. *Assume that $f : \mathbb{R} \times \mathbb{X} \mapsto \mathbb{X}$ satisfies a Lipschitz condition in x uniformly in t, that is,*

$$\|f(t,x) - f(t,y)\| \leq L\|x-y\|, \quad \forall x, y \in \mathbb{X},$$

where $L < \dfrac{|\omega|}{K}$. Let also f be almost automorphic in t for each $x \in \mathbb{X}$. Then equation (4.6) has a unique almost automorphic mild solution.

Proof. Let x be a mild solution of Eq. (4.6). It is continuous and satisfies the integral equation

$$x(t) = T(t-a)x(a) + \int_a^t T(t-s)f(s, x(s))ds, \quad \forall a \in \mathbb{R}, \quad \forall t \geq a.$$

Consider $\int_a^t T(t-s)f(s,x(s))ds$ and the nonlinear operator $G : AA(\mathbb{X}) \mapsto AA(\mathbb{X})$ given by

$$(G\phi)(t) := \int_{-\infty}^t T(t-s)f(s, \phi(s))ds.$$

In view of Theorem 2.2.6 in [N'Guérékata (79)], if $\phi \in AA(\mathbb{X})$, then $G\phi \in AA(\mathbb{X})$, so G is well-defined.

Now for $\phi_1, \ \phi_2 \in AA(\mathbb{X})$, we have:

$$\begin{aligned}
\|G\phi_1 - G\phi_2\|_\infty &= \sup_{t\in\mathbb{R}} \| \int_{-\infty}^t T(t-s)\{f(s,\phi_1(s)) - f(s,\phi_2(s))\}ds\| \\
&\leq \sup_{t\in\mathbb{R}} \int_{-\infty}^t \|T(t-s)\|_{B(\mathbb{X})} L \|\phi_1(s) - \phi_2(s)\| ds \\
&\leq L\|\phi_1 - \phi_2\|_\infty . \sup_{t\in\mathbb{R}} \int_{-\infty}^t \|T(t-s)\|_{B(\mathbb{X})} ds \\
&\leq L\|\phi_1 - \phi_2\|_\infty . \sup_{t\in\mathbb{R}} \int_{-\infty}^t K e^{\omega(t-s)} ds \\
&= \frac{LK}{|\omega|} \|\phi_1 - \phi_2\|_\infty .
\end{aligned}$$

So

$$\|G\phi_1 - G\phi_2\|_\infty \leq \frac{LK}{|\omega|} \|\phi_1 - \phi_2\|_\infty,$$

which proves that G is continuous. And since $\dfrac{LK}{|\omega|} < 1$, then G is a contraction. So there exists a unique $u \in AA(\mathbb{X})$, such that $Gu = u$, that is $u(t) = \int_{-\infty}^t T(t-s)f(s,u(s))ds$.

If we let $u(a) = \int_{-\infty}^a T(a-s)f(s,u(s))ds$, then

$$T(t-a)u(a) = \int_{-\infty}^a T(t-s)f(s,u(s))ds.$$

But for $t \geq a$,

$$\begin{aligned}
\int_a^t T(t-s)f(s,u(s))ds &= \int_{-\infty}^t T(t-s)f(s,u(s))ds \\
&\quad - \int_{-\infty}^a T(t-s)f(s,u(s))ds \\
&= u(t) - T(t-a)u(a).
\end{aligned}$$

So $u(t) = T(t-a)u(a) + \int_{-\infty}^t T(t-s)f(s,u(s))ds$ is a mild solution of equation (4.6) and $u \in AA(\mathbb{X})$. The proof is now complete. $\square$

4.4 Method of Sums of Commuting Operators and Almost Automorphic Functions

In this section we will extend the method of sums of commuting operators in the previous chapter to almost automorphic solutions of inhomogeneous linear evolution equations of the form

$$\frac{du}{dt} = Au + f(t), \tag{4.9}$$

where A is an (unbounded) linear operator which generates a holomorphic semigroup of linear operators on a Banach space $\mathbb{X}$ and $f \in AA(\mathbb{X})$.

The main difficulty that arises here is concerned with the non-uniform continuity of almost automorphic functions. This implies the non-strong continuity of translation semigroup in the functions space $AA(\mathbb{X})$. Therefore, many elegant proofs using semigroup theory fail. To overcome this difficulty it is appropriate to use the concept of uniform spectrum defined in Chapter 1.

For any closed subset $\Lambda \subset \mathbb{R}$ we denote

$$AA_\Lambda(\mathbb{X}) := \{u \in AA(\mathbb{X}) : sp_u(u) \subset \Lambda\}.$$

By the basic properties of uniform spectra of functions, $AA_\Lambda(\mathbb{X})$ is a closed subspace of $BC(\mathbb{R}, \mathbb{X})$. Below we denote $\mathcal{D}_\Lambda$ the part of the differential operator d/dt in $AA_\Lambda(\mathbb{X})$. We have the following:

Lemma 4.1. *Under the above notations and assumptions we have*

$$\sigma(\mathcal{D}_\Lambda) = i\Lambda. \tag{4.10}$$

Proof. The proof can be taken from the one of Lemma 1.2. The details are left to the reader. □

As a *standing assumption* in the remaining part of the section we always assume that A is the infinitesimal generator of an analytic semigroup of linear operators on $\mathbb{X}$.

Let Λ be a closed subset of $\mathbb{R}$. We first consider the operator $\mathcal{A}_\Lambda$ of multiplication by A and the differential operator d/dt on the function space $AA_\Lambda(\mathbb{X})$.

By definition the operator $\mathcal{A}_\Lambda$ of multiplication by A is defined on $D(\mathcal{A}_\Lambda) := \{g \in AA_\Lambda(\mathbb{X}) : g(t) \in D(A)\ \forall t \in \mathbb{R}, Ag(\cdot) \in AA_\Lambda(\mathbb{X})\}$, and $\mathcal{A}g := Ag(\cdot)$ for all $g \in D(\mathcal{A}_\Lambda)$.

Lemma 4.2. *Assume that $\Lambda \subset \mathbb{R}$ is closed. Then the operator $\mathcal{A}_\Lambda$ of multiplication by A in $AA_\Lambda(\mathbb{X})$ is the infinitesimal generator of an analytic C_0-semigroup on $AA_\Lambda(\mathbb{X})$.*

Proof. We will prove that $\mathcal{A}_\Lambda$ is a sectorial operator on $AA_\Lambda(\mathbb{X})$. In fact, first we check that $\mathcal{A}_\Lambda$ is densely defined. Consider the semigroup $\mathcal{T}_\Lambda(t)$ of operators of multiplication by $T(t)$ on $AA_\Lambda(\mathbb{X})$. We now show that it is strongly continuous. Indeed, suppose that $g \in AA_\Lambda(\mathbb{X})$, since $R(g)$ is relatively compact we see that the map $[0,1] \times \overline{R(g)} \ni (t,x) \mapsto T(t)x \in \mathbb{X}$ is uniformly continuous. Hence,

$$\sup_{s\in\mathbb{R}} \|T(t)g(s) - g(s)\| \to 0$$

as $t \to 0$, i.e., the $\mathcal{T}_\Lambda(t)$ is strongly continuous. By definition, $g \in D(\mathcal{A}_\Lambda)$ if and only if $g(s) \in D(A)$, $\forall s \in \mathbb{R}$ and $Ag(\cdot) \in AA_\Lambda(\mathbb{X})$. Thus,

$$\frac{T(t)g(s) - g(s)}{t} = \frac{1}{t}\int_0^t T(\xi)Ag(s)d\xi, \quad \forall t \geq 0, s \in \mathbb{R}.$$

Therefore,

$$\lim_{t\to 0^+} \sup_{s\in\mathbb{R}} \| \frac{T(t)g(s) - g(s)}{t} - \frac{1}{t}\int_0^t T(\xi)Ag(s)d\xi\| = 0,$$

i.e., g is in $D(G)$, where G is the generator of $\mathcal{T}_\Lambda(t)$ and $\mathcal{A}_\Lambda g = Gg$. Conversely, we can easily show that $G \subset \mathcal{A}_\Lambda$.

Now it suffices to prove that $\sigma(\mathcal{A}_\Lambda) \subset \sigma(A)$ to claim that $\mathcal{A}_\Lambda$ is a sectorial operator. In fact, let $\mu \in \rho(A)$. To prove that $\mu \in \rho(\mathcal{A}_\Lambda)$ we show that for each $h \in AA_\Lambda(\mathbb{X})$ the equation $\mu g - \mathcal{A}_\Lambda g = h$ has a unique solution in $AA_\Lambda(\mathbb{X})$.

But this follows from the fact that $(\mu - \mathcal{A}_\Lambda)^{-1}h(\cdot) \in AA_\Lambda(\mathbb{X})$ and that the equation

$$\mu x - Ax = y$$

has a unique solution x in $\mathbb{X}$ for any $y \in \mathbb{X}$. □

Theorem 4.8. *Let A be the generator of an analytic semigroup. Then the operator $\mathcal{A}_\Lambda$ of multiplication by A and the differential operator $\mathcal{D}_\Lambda$ on $AA_\Lambda(\mathbb{X})$ are commuting and satisfy condition P (for the definition see the Appendix).*

Proof. By the above lemma the operator $\mathcal{A}_\Lambda$ is sectorial. It suffices to show that it commutes with the differential operator $\mathcal{D}_\Lambda$. In fact, since $1 \in \rho(\mathcal{D}_\Lambda)$ we will prove that

$$R(1, \mathcal{D}_\Lambda)R(\omega, \mathcal{A}_\Lambda) = R(\omega, \mathcal{A}_\Lambda)R(1, \mathcal{D}_\Lambda), \tag{4.11}$$

for sufficiently large real ω.

Since $\mathcal{A}_\Lambda$ generates the semigroup $\mathcal{T}_\Lambda(t)$, using well-known facts from the semigroup theory the above indentity for sufficiently large ω is equivalent to the following

$$R(1, \mathcal{D}_\Lambda)\int_0^\infty e^{-\omega t}\mathcal{T}_\Lambda(t)dt = \int_0^\infty e^{-\omega t}\mathcal{T}_\Lambda(t)dt R(1, \mathcal{D}_\Lambda). \tag{4.12}$$

In turn, (4.12) follows from the following

$$R(1, \mathcal{D}_\Lambda)\mathcal{T}_\Lambda(\tau) = \mathcal{T}_\Lambda(\tau)R(1, \mathcal{D}_\Lambda), \quad \forall \tau \geq 0. \tag{4.13}$$

which is obvious. □

So, by the spectral properties of sums of commuting operators, we have

Corollary 4.2. *If $\sigma(A) \cap i\Lambda = \oslash$, then for every $f \in AA_\Lambda(\mathbb{X})$ there exists a unique $u \in AA_\Lambda(\mathbb{X})$ such that*

$$\overline{\mathcal{A}_\Lambda + \mathcal{D}_\Lambda} u = f.$$

Proof. Since $\mathcal{A}_\Lambda$ and $\mathcal{D}_\Lambda$ commute and satisfy Condition P, the sum $\mathcal{A}_\Lambda + \mathcal{D}_\Lambda$ is closable (denote its closure by $\overline{\mathcal{A}_\Lambda + \mathcal{D}_\Lambda}$).

From $\sigma(A) \cap i\Lambda = \oslash$ and Theorem 1.11 in Appendix, it turns out that $0 \in \rho(\overline{\mathcal{A}_\Lambda + \mathcal{D}_\Lambda})$.

Therefore for every $f \in AA_\Lambda(\mathbb{X})$ there exists a unique $u \in D(\overline{\mathcal{A}_\Lambda + \mathcal{D}_\Lambda})$ such that

$$\overline{\mathcal{A}_\Lambda + \mathcal{D}_\Lambda} u = f.$$

□

Now our remaining task is just to explain what the above closure means. More precisely, we will relate it with the notion of mild solutions to evolution equations.

Lemma 4.3. *Let $u, f \in AA(\mathbb{X})$. If $u \in D(\overline{\mathcal{A}_\Lambda + \mathcal{D}_\Lambda})$ and $\overline{\mathcal{A}_\Lambda + \mathcal{D}_\Lambda} u = f$, then u is a mild solution of Eq. (4.11).*

Proof. This lemma follows immediately from the following:

For every $u \in AA(\mathbb{X})$ we say that it belongs to $D(L)$ of an operator L acting on $AA(\mathbb{X})$ if there is a function $f \in AA(\mathbb{X})$ such that

$$u(t) = T(t-s)u(s) + \int_s^t T(t-\xi)f(\xi)d\xi, \quad \forall t \geq s, t, s \in \mathbb{R}. \tag{4.14}$$

By a similar argument as in the proof of Lemma 3.1 in [Murakami, Naito and Minh (71)] we can prove that L is a closed single-valued linear operator acting on $AA(\mathbb{X})$ which is an extension of $\mathcal{A}_\Lambda + \mathcal{D}_\Lambda$.

Thus, L is an extension of $\overline{\mathcal{A}_\Lambda + \mathcal{D}_\Lambda}$. This yields that u is a mild solution of Eq. (4.11). □

As an immediate consequence of the above argument we have:

Theorem 4.9. *Let A be the generator of an analytic semigroup and let Λ be a closed subset of $\mathbb{R}$.*

Then it is necessary and sufficient for each $f \in AA_\Lambda(\mathbb{X})$ there exists a unique mild solution $u \in AA_\Lambda(\mathbb{X})$ to Eq. (4.11) that the condition $\sigma(A) \cap i\Lambda = \oslash$ holds.

Proof. The sufficiency follows from the above argument. The necessity can be shown as follows:

For every $\xi \in \Lambda$, obviously that the function $h : \mathbb{R} \ni t \mapsto ae^{i\xi t}$ is in $AA_\Lambda(X)$, where $a \in \mathbb{X}$ is any given element.

By assumption, there is a unique $g \in D(\mathcal{A}_\Lambda)$ such that $i\xi g(t) - Ag(t) = h(t)$ for all $t \in \mathbb{R}$.

Following the argument in the proof of Theorem 3.4 one can easily show that $g(t)$ is of the form $be^{i\xi t}$. Hence, b is the unique solution of the equation $i\xi b - Ab = a$.

That is $i\xi \notin \sigma(\mathcal{A}_\Lambda)$, so $i\Lambda \cap \sigma(\mathcal{A}_\Lambda) = \oslash$. □

Corollary 4.3. *Let A be the generator of an analytic semigroup such that $\sigma(A) \cap i\ sp_u(f) = \oslash$.*

Then Eq. (4.11) has a unique almost automorphic mild solution w such that $sp_u(w) \subset sp_u(f)$.

Proof. Set $\Lambda = sp_u(f)$. Then by the above argument we get the theorem. □

Remark 4.3. We notice that all results stated above for almost automorphic solutions hold true for compact almost automorphic solutions if the assumption on the almost automorphy of f is replaced by the compact almost automorphy of f. Details of the proofs are left to the reader.

4.5 Almost Automorphic Solutions of Second Order Evolution Equations

In this section we deal with the existence of almost automorphic mild solutions to second order evolution equations of the form

$$\frac{d^2u}{dt^2} = Au + f(t), \tag{4.15}$$

where A is an (unbounded) linear operator which generates a holomorphic semigroup of linear operators on a Banach space $\mathbb{X}$ and $f \in AA(\mathbb{X})$.

We will use the method of sums of commuting operators, a method due to Murakami, Naito and Minh [Murakami, Naito and Minh (71)], to study the existence of almost periodic solutions is . This method works well with the problem of finding almost automorphic solutions of first order evolution equations as shown in the previous section 4.4.

We will show that the method of sums is still useful for second order evolution equations. Our main results are Theorems 4.9, 4.11. Notice that in [Schweiker (94)], [Schuler and Vu (95)], similar results for $f \in BUC(\mathbb{R},\mathbb{X})$ were proved using an operator equation. This method that needs the uniform continuity of f is inapplicable to the problem we are considering due to the fact that an almost automorphic function may not be uniformly continuous.

Notation. Throughout the section, $\mathbb{R}$, $\mathbb{C}$, $\mathbb{X}$ stand for the sets of all real, all complex numbers and a complex Banach space, respectively; $L(\mathbb{X})$, $BC(\mathbb{R},\mathbb{X})$, $BUC(\mathbb{R},\mathbb{X})$ denote the spaces of all linear bounded operators on $\mathbb{X}$, all $\mathbb{X}$-valued bounded continuous functions, all $\mathbb{X}$-valued bounded uniformly continuous with sup-norm, respectively. The translation group in $BC(\mathbb{R},\mathbb{X})$ is denoted by $(S(t))_{t\in\mathbb{R}}$ which is strongly continuous in $BUC(\mathbb{R},\mathbb{X})$ whose infinitesimal generator is the differential operator d/dt. For a linear operator A, we denote by $D(A)$, $\sigma(A)$ and $\rho(A)$ the domain, spectrum and resolvent set of A, respectively. If Y is a metric space and B is a subset of Y, then $\bar{B}$ denotes its closure in Y. In this section by the notion of sectorial operators means the one defined in [Pazy (90)]. The notion of closure of an operator is referred to the one defined in [Davies (27)].

4.5.1 *Mild Solutions of Inhomogeneous Second Order Equations*

4.5.1.1 *Mild Solutions*

Let A be any closed linear operator on a Banach space $\mathbb{X}$. We now define the concept of mild solutions on $\mathbb{R}$ to Eq. (4.15).

Definition 4.1. (See p. 374 in [Arendt and Batty (5)]) A continuous $\mathbb{X}$-valued function u is called a *mild solution on* $\mathbb{R}$ of Eq. (4.15) if

$$\int_0^t (t-s)u(s)ds \in D(A) \quad \forall t \in \mathbb{R} \tag{4.16}$$

and

$$u(t) = x + ty + A\int_0^t (t-s)u(s)ds + \int_0^t (t-s)f(s)ds \quad (t \in \mathbb{R}) \tag{4.17}$$

for some fixed $x, y \in \mathbb{X}$. If $u \in C^2(\mathbb{R},\mathbb{X})$, $u(t) \in D(A), \forall t \in \mathbb{R}$ and Eq.(4.15) holds for all $t \in \mathbb{R}$ we say that u is a *classical solution* to Eq.(4.15). It is easily seen that if $u \in C(\mathbb{R},\mathbb{X})$ is a classical solution of Eq. (4.15), then it is a mild solution of Eq. (4.15).

4.5.1.2 *Mild Solutions and Weak solutions*

Definition 4.2. A function $u \in BC(\mathbb{R},\mathbb{X})$ is said to be a weak solution to Eq. (4.15) if there is a sequence of $f_n \in BC(\mathbb{R},\mathbb{X})$ and a sequence of classical solutions $u_n \in BC(\mathbb{R},\mathbb{X})$ of Eq. (4.15) with f replaced by f_n such that $f_n \to f$ and $u_n \to u$ in the sup-norm topology of $BC(\mathbb{R},\mathbb{X})$.

Definition 4.3. We define an operator $\mathcal{L}$ on $BC(\mathbb{R},\mathbb{X})$ with domain $D(\mathcal{L})$ consisting of all $u \in BC(\mathbb{R},\mathbb{X})$ such that there exists at least a function $f \in BC(\mathbb{R},\mathbb{X})$ for which u is a mild solution of Eq. (4.15), that is, (4.16) and (4.17) hold.

Lemma 4.4. *If A is a closed linear operator, then $\mathcal{L}$ is a single-valued closed linear operator $BC(\mathbb{R},\mathbb{X}) \supset D(\mathcal{L}) \to BC(\mathbb{R},\mathbb{X})$.*

Proof. First we show that $\mathcal{L}$ is a single-valued linear operator. For this it suffices to show that if $u(t) \equiv 0$ is a mild solution of Eq. (4.15), then $f(t) \equiv 0$. In fact, by taking $t = 0$ we can see that in Eq. (4.17), $x = 0$. Hence, we have

$$0 = ty + \int_0^t (t-s)f(s)ds \quad (t \in \mathbb{R}). \tag{4.18}$$

Differentiating (4.18) twice, we have $f(t) = 0$ for all $t \in \mathbb{R}$. Therefore, the operator $\mathcal{L}$ is a single-valued linear operator.

Next, we show its closedness. Let u_n be in $D(\mathcal{L})$ that are mild solutions of Eq. (4.15) with f replaced by $f_n \in BC(\mathbb{R}, \mathbb{X})$ such that $u_n \to u \in BC(\mathbb{R}, \mathbb{X})$ and $f_n \to f \in BC(\mathbb{R}, \mathbb{X})$. We have to prove that u is a mild solution to Eq. (4.15). Indeed, by our assumption, we have

$$u_n(t) = x + ty + A\int_0^t (t-s)u_n(s)ds + \int_0^t (t-s)f_n(s)ds \quad (t \in \mathbb{R}).$$

Since A is closed and $u_n, f_n \in BC(\mathbb{R}, \mathbb{X})$, for a fixed $t \in \mathbb{R}$, letting $n \to \infty$ we have $\int_0^t (t-s)u(s)ds \in D(A)$, and $A\int_0^t (t-s)u_n(s)ds \to A\int_0^t (t-s)u(s)ds$. Therefore,

$$u(t) = x + ty + A\int_0^t (t-s)u(s)ds + \int_0^t (t-s)f(s)ds \quad (t \in \mathbb{R}).$$

Thus, by definition, u is a mild solution of Eq. (4.15). □

Remark 4.4. By a similar argument, we can easily show that the part of $\mathcal{L}$ on AA_Λ (here Λ is a closed subspace of $\mathbb{R}$) that is denoted by $\mathcal{L}_\Lambda$ is a closed linear operator.

Proposition 4.1. *Let A be a closed linear operator. Then every weak solution of Eq. (4.15) is a mild solution.*

Proof. The proof is obvious in view of the above lemma and the remark that classical solutions are mild solutions. □

4.5.2 *Operators $\mathcal{A}$*

Let Λ be a closed subset of $\mathbb{R}$. We first consider the operator $\mathcal{A}_\Lambda$ of multiplication by A and the differential operator d/dt on the function space $AA_\Lambda(\mathbb{X})$. By definition the operator $\mathcal{A}_\Lambda$ of multiplication by A is defined on $D(\mathcal{A}_\Lambda) := \{g \in AA_\Lambda(\mathbb{X}) : g(t) \in D(A)\ \forall t \in \mathbb{R}, Ag(\cdot) \in AA_\Lambda(\mathbb{X})\}$, and $\mathcal{A}g := Ag(\cdot)$ for all $g \in D(\mathcal{A}_\Lambda)$.

In the following result we assume that Λ is a closed subset of the real line and the second order differential operator $\mathcal{D}^2_\Lambda$ defined in $AA_\Lambda(\mathbb{X})$ with domain $D(\mathcal{D}^2_\Lambda)$ consisting of all functions $u \in AA_\Lambda(\mathbb{X})$ that are of class C^2 such that $d^2u/dt^2 \in AA_\Lambda(\mathbb{X})$. Let us define $-\Lambda^2 := \{\xi \in \mathbb{R} | \xi = -\lambda^2$, for some $\lambda \in \Lambda\}$.

Proposition 4.2. *With the above notations the following assertions hold true*

$$\sigma(\mathcal{D}^2_\Lambda) = -\Lambda^2. \tag{4.19}$$

Proof. We associate with the equation

$$\frac{d^2u}{dt^2} = \mu u + f(t), f \in AA_\Lambda(\mathbb{X}) \tag{4.20}$$

the following first order equation

$$\left\{ x_1' = x_2 x_2' = \mu x_1 + f(t), \quad f \in AA_\Lambda(\mathbb{X}). \right. \tag{4.21}$$

It is easily seen from the theory of ODE that the solvability of these equations in $BC(\mathbb{R}, \mathbb{X})$ are equivalent. As shown in Theorem 4.10, if $f \in AA_\Lambda(\mathbb{X})$ and

$$i\Lambda \cap \sigma(I(\mu)) = \oslash, \tag{4.22}$$

where $I(\mu)$ denotes the operator matrix associated with Eq.(4.21). A simple computation shows that $\sigma(I(\mu))$ consists of all solutions to the equation $t^2 - \mu = 0$. Thus,

$$\sigma(\mathcal{D}^2_\Lambda) \subset \{\mu \in \mathbb{C} : \mu = -\lambda^2 \text{ for some } \lambda \in \Lambda\}.$$

Hence $\sigma(\mathcal{D}^2_\Lambda) \subset -\Lambda^2$.

On the other hand, let $\mu \in \Lambda$. Then $g(\cdot) := xe^{i\mu\cdot} \in AA_\Lambda(\mathbb{X})$. Obviously, $\mathcal{D}^2_\Lambda g = -\mu^2 g$ and thus, $-\mu^2 \in \sigma(\mathcal{D}^2_\Lambda)$. Therefore, $\sigma(\mathcal{D}^2_\Lambda) \supset -\Lambda^2$. That is, the proposition is proven. □

Proposition 4.3. *Let Λ be a closed subset of the real line and let $\Sigma(\varepsilon)$ denote the set $\{z \in \mathbb{C} | z \neq 0, |\arg z| < \pi - \varepsilon\}$, where ε is a given small positive number. Then the resolvent $R(\lambda, \mathcal{D}^2)$ of the operator $\mathcal{D}^2$ in $AA_\Lambda(\mathbb{X})$ satisfies*

$$\|R(\lambda, \mathcal{D}^2)\| \leq \frac{M}{|\lambda|}, \quad \forall \lambda \in \Sigma(\varepsilon), \tag{4.23}$$

where M is a positive constant depending only on ε.

Proof. Consider the equation

$$x''(t) - \lambda x(t) = f(t), \quad t \in \mathbb{R}, \tag{4.24}$$

for every given $\lambda \in \Sigma(\varepsilon)$ and $f \in AA_\Lambda(\mathbb{X})$. As shown in the previous proposition, there exists a unique function $x_f \in AA_\Lambda(\mathbb{X})$ that satisfies the above equation. For $\lambda \in \Sigma(\varepsilon)$ there are exactly two distinct solutions $\lambda_1, \lambda_2 = -\lambda_1$ of the equation $x^2 - \lambda = 0$. Without loss of generality, we assume that $\Re e\lambda_1 < 0$. It is easy to check that the unique bounded solution to Eq. (4.24) is given by

$$x_f(t) = \frac{1}{2\lambda_1}\left(\int_{-\infty}^{t} e^{\lambda_1(t-\xi)} f(\xi) d\xi + \int_{t}^{\infty} e^{-\lambda_1(t-\xi)} f(\xi) d\xi\right) \tag{4.25}$$

that is in $AA_\Lambda(\mathbb{X})$. By definition, $x_f = -R(\lambda, \mathcal{D}^2)f$. Obviously, since $\lambda \in \Sigma(\varepsilon)$, we have

$$\|x_f\| \leq \frac{1}{2|\lambda_1|} \cdot \frac{2}{\Re e\lambda_1} \leq \frac{M}{|\lambda|},$$

where M is a positive constant depending on ε. □

Proposition 4.4. *Let A be a closed linear operator such that $\sigma(A) \cap -\Lambda^2 = \oslash$. Then for every $f \in BC(\mathbb{R}, \mathbb{X})$ with $sp(f) \subset \Lambda$, Eq. (4.15) has at most one mild solution $u \in BC(\mathbb{R}, \mathbb{X})$.*

Proof. Since the operator $\mathcal{L}$ is a closed linear operator, it suffices to show that if $u \in BC(\mathbb{R}, \mathbb{X})$ is a mild solution of Eq. (4.15) with f replaced by 0, then $sp(u) = \oslash$. This can be checked directly as follows. Assume that u is a solution of Eq. (4.17) with f replaced by 0.

Taking Carleman-Laplace transforms of both sides of Eq. (4.17) we have

$$\hat{u}(\lambda) = \frac{x}{\lambda} + \frac{y}{\lambda^2} + A\frac{\hat{u}(\lambda)}{\lambda^2} \quad (\Re\lambda \neq 0).$$

Therefore,

$$(\lambda^2 - A)\hat{u}(\lambda) = \lambda x + y, \quad (\Re\lambda \neq 0). \tag{4.26}$$

If $\xi \in \mathbb{R}$ and $\xi \notin \sigma(A)$, then $(\xi^2 - A)$ is invertible and $(\lambda^2 - A)^{-1}$ is holomorphic in a small neighborhood of ξ. Therefore, $\hat{u}(\lambda)$ has a holomorphic extension to a neighborhood of ξ, so $sp(u) \subset \mathbb{R} \cap \sigma(A)$. On the other hand, by the assumption that $sp(u) \subset -\Lambda^2$ and $-\Lambda^2 \cap \sigma(A) = \oslash$, we have $sp(u) = \oslash$. And hence, $u = 0$. □

Lemma 4.5. *Let A be the generator of an analytic semigroup. Then the operator $\mathcal{A}_\Lambda$ of multiplication by A and the differential operator $\mathcal{D}^2_\Lambda$ on $AA_\Lambda(\mathbb{X})$ are commuting and satisfy condition P of Definition 1.11.*

Proof. Let us denote by $T(t)$ the semigroup generated by A. As shown in the previous section 4.4, the operator $\mathcal{A}$ of multiplication by A generates an analytic semigroup $\mathcal{T}(t)$ of multiplication by $T(t)$ on $AA_\Lambda(\mathbb{X})$. So, it is sufficient to prove that the semigroup $\mathcal{T}(t)$ commutes with $\mathcal{D}^2$, that is $\mathcal{T}(t)D(\mathcal{D}^2) \subset D(\mathcal{D}^2)$ and

$$\mathcal{T}(t)\mathcal{D}^2 x = \mathcal{D}^2\mathcal{T}(t)x, \quad \forall x \in D(\mathcal{D}^2),\ t \geq 0.$$

By the definition of the operator $\mathcal{T}(t)$ of multiplication by $T(t)$, the above claim is obvious. The lemma now follows from the above propositions. □

So, by the spectral properties of sums of commuting operators, we have

Corollary 4.4. *If A generates an analytic semigroup with $\sigma(A) \cap -\Lambda^2 = \oslash$, then for every $f \in AA_\Lambda(\mathbb{X})$ there exists a unique $u \in AA_\Lambda(\mathbb{X})$ such that*

$$\overline{\mathcal{D}_\Lambda^2 - \mathcal{A}_\Lambda} u = f.$$

Proof. Since $\mathcal{A}_\Lambda$ and $-\mathcal{D}_\Lambda^2$ commute and satisfy Condition P, the sum $\mathcal{D}_\Lambda^2 - \mathcal{A}_\Lambda$ is closable (denote its closure by $\overline{\mathcal{D}_\Lambda^2 - \mathcal{A}_\Lambda}$). From $\sigma(A) \cap -\Lambda^2 = \oslash$ and Theorem 1.11 in Appendix, it turns out that $0 \in \rho(\overline{\mathcal{D}_\Lambda^2 - \mathcal{A}_\Lambda})$. Therefore, for every $f \in AA_\Lambda(\mathbb{X})$ there exists a unique $u \in D(\overline{\mathcal{D}_\Lambda^2 - \mathcal{A}_\Lambda})$ such that

$$\overline{\mathcal{D}_\Lambda^2 - \mathcal{A}_\Lambda} u = f. \qquad \square$$

Now our remaining task is just to explain what the above closure means. More precisely, we will relate it with the notion of mild solutions to evolution equations.

Lemma 4.6. *Let $u, f \in AA(\mathbb{X})$. If $u \in D(\overline{\mathcal{D}_\Lambda^2 - \mathcal{A}_\Lambda})$ and $\overline{\mathcal{D}_\Lambda^2 - \mathcal{A}_\Lambda} u = f$, then u is a mild solution of Eq. (4.15).*

Proof. The lemma follows from the fact that weak solutions are mild solutions. $\square$

As an immediate consequence of the above argument we have:

Theorem 4.10. *Let A be the generator of an analytic semigroup and let Λ be a closed subset of $\mathbb{R}$. Then it is necessary and sufficient for each $f \in AA_\Lambda(\mathbb{X})$ there exists a unique mild solution $u \in AA_\Lambda(\mathbb{X})$ to Eq. (4.15) that the condition $\sigma(A) \cap -\Lambda^2 = \oslash$ holds.*

Proof. The sufficiency follows from the above argument. The necessity can be shown as follows: For every $\xi \in \Lambda$, obviously that the function $h : \mathbb{R} \ni t \mapsto ae^{i\xi t}$ is in $AA_\Lambda(X)$, where $a \in \mathbb{X}$ is any given element. By assumption, there is a unique $g \in D(\mathcal{A}_\Lambda)$ such that $-\xi g(t) - Ag(t) = h(t)$ for all $t \in \mathbb{R}$. Following the argument in p. 252 of [Murakami, Naito and Minh (71)] one can easily show that $g(t)$ is of the form $be^{i\xi t}$. Hence, b is the unique solution of the equation $-\xi b - Ab = a$. That is $-\xi \notin \sigma(\mathcal{A}_\Lambda)$, so $-\Lambda \cap \sigma(\mathcal{A}_\Lambda) = \oslash$. $\square$

Corollary 4.5. *Let A be the generator of an analytic semigroup such that $\sigma(A) \cap [-sp_u(f)]^2 = \oslash$. Then Eq. (4.15) has a unique almost automorphic mild solution w such that $sp_u(w) \subset sp_u(f)$.*

Proof. Set $\Lambda = sp_u(f)$. Then by the above argument we get the theorem. $\square$

Remark 4.5. We notice that all results stated above for almost automorphic solutions hold true for compact almost automorphic solutions if the assumption on the almost automorphy of f is replaced by the compact almost automorphy of f. Details of the proofs are left to the reader.

4.5.3 *Nonlinear Equations*

In this subsection we consider nonlinear equations of the form

$$\frac{d^2}{dt^2}u(t) = Au(t) + F(t, u(t)), \quad t \in \mathbb{R}, \tag{4.27}$$

where $F : \mathbb{R} \times \mathbb{X} \to \mathbb{X}$ is assumed to satisfies the following conditions: $F(t, x)$ is almost automorphic in t for every fixed $x \in \mathbb{X}$ and F is continuous jointly in (t, x), Lipschtz in x uniformly in t, that is, there is a positive number ε independent of t, x such that

$$\|F(t, x) - F(t, y)\| \leq \varepsilon \|x - y\|, \quad \forall x, y \in \mathbb{X}. \tag{4.28}$$

We will say that *Condition H* holds for F in $AA_\Lambda(\mathbb{X})$ if for a closed subset $\Lambda \subset \mathbb{R}$ the Nemystky operator $\mathcal{F}$, defined by

$$\mathcal{F}g(t) := F(t, g(t)), \quad \forall g \in AA_\Lambda(\mathbb{X}), t \in \mathbb{R}$$

is an operator $AA_\Lambda(\mathbb{X}) \to AA_\Lambda(\mathbb{X})$.

Lemma 4.7. *If $F(t, x)$ is almost automorphic in t for every fixed $x \in \mathbb{X}$, F is continuous jointly in (t, x) and satisfies (4.28), then the Nemystky operator $\mathcal{F}$ is a continuous operator acting on $AA(\mathbb{X})$, that is, condition H holds for F on $AA_\mathbb{R}(\mathbb{X})$.*

Proof. For the proof see Theorem 2.2.6 in [N'Guérékata (79)]. $\square$

Remark 4.6. In general, we are still puzzled with conditions on F and Λ such that condition H holds for F on $AA_\Lambda(\mathbb{X})$. In the linear case, if $F(t, x) = L(t)x$ is T-periodic in t and $\Lambda = \{\xi \in \mathbb{R} | e^{i\xi} \in G\}$, where G is a closed subset of the unit circle, then condition H holds for F and Λ (see e.g. [Batty, Hutter and Räbiger (14)]). In the general nonlinear case, this question is still open.

Definition 4.4. A continuous $\mathbb{X}$-valued function u is called a *mild solution on* $\mathbb{R}$ of (4.27) if

$$\int_0^t (t - s)u(s)ds \in D(A), \quad \forall t \in \mathbb{R} \tag{4.29}$$

and

$$u(t) = x + ty + A\int_0^t (t-s)u(s)ds + \int_0^t (t-s)f(s,u(s))ds \quad (t \in \mathbb{R}) \tag{4.30}$$

for some fixed $x, y \in \mathbb{X}$. If $u \in C^2(\mathbb{R}, \mathbb{X})$, $u(t) \in D(A), \forall t \in \mathbb{R}$ and Eq.(4.27) holds for all $t \in \mathbb{R}$ we say that u is a *classical solution* to Eq.(4.27). It is easily seen that if $u \in C(\mathbb{R}, \mathbb{X})$ is a classical solution of Eq. (4.27), then it is a mild solution of Eq. (4.27).

The following is a main result of this subsection:

Theorem 4.11. *Let A be the generator of an analytic semigroup, and let condition H hold for F on $AA_\Lambda(\mathbb{X})$. Furthermore, we assume that $\sigma(A) \cap -\Lambda^2 = \oslash$ and the Lipshitz coefficient ε in (4.28) is sufficiently small. Then, Eq. (4.27) has a unique almost automorphic mild solution $u \in AA_\Lambda(\mathbb{X})$.*

Proof. First we consider the operator $\mathcal{L}_\Lambda$ defined in Definition 4.3 and the remark that follows. Under the assumptions, by Theorem 4.10, this operator is invertible on $AA_\Lambda(\mathbb{X})$. Next, we consider the operator $\mathcal{L}_\Lambda - \mathcal{F}$. This operator may be seen as a nonlinear perturbation of $\mathcal{L}_\Lambda$. Since $\mathcal{L}_\Lambda$ is a closed linear operator, if ε is sufficiently small, and if we consider $D(\mathcal{L}_\Lambda) \subset AA_\Lambda(\mathbb{X})$ with the graph norm of $\mathcal{L}_\Lambda$, then by the Inverse Function Theorem, the operator $\mathcal{L}_\Lambda - \mathcal{F}$ is an invertible Lipschitz operator from $D(\mathcal{L}_\Lambda)$ onto $AA_\Lambda(\mathbb{X})$. Hence, Eq. (4.27) has a unique almost automorphic mild solution. □

4.6 The Equations x'=f(t,x)

In this section we are going to study the existence of almost automorphic solutions of nonlinear differential equations and characterize some topological properties of the almost automorphic solutions sets . Thus we consider in a Banach space $\mathbb{X}$ the Cauchy problem **CP**

$$x' = f(t,x), \qquad x(0) = x_0, \quad t \in \mathbb{R}$$

where $x_0 \in \mathbb{X}$ and $f : \mathbb{R} \times \mathbb{X} \to \mathbb{X}$.

Let us recall the following:

Definition 4.5. Let X, Y be metric spaces, $f : X \to Y$ be a continuous function and let $y \in Y$. f is said to be proper at the point y provided that there exists $\varepsilon > 0$ such that for any compact set $K \subset B(y, \varepsilon)$ the set

$f^{-1}(K)$ is compact, where $B(y, \varepsilon)$ is the open ball in Y of center y and radius ε.

If $f^{-1}(K)$ is compact for any compact $K \subset Y$, then f is called proper.

Now we denote by α the Kuratowski measure of noncompactness and assume the following:

1^0 f is a continuous mapping such that

$$\|f(s,x)\| \le \phi(s) \quad \text{and} \quad \int_{-\infty}^{+\infty} \phi(s)ds < +\infty;$$

2^0 for each bounded subset $M \subset \mathbb{X}$ and for each interval $[\min(0,a), \max(0,a)]$, there exists a continuous nondecreasing function $h_{M,a} : \mathbb{R}_+ \to \mathbb{R}_+$ such that the inequality

$$u(t) \le \int_{\min(0,t)}^{\max(0,t)} h_{M,a}(u(s))ds, \quad \text{for } t \in [\min(0,a), \max(0,a)],$$

has only a trivial solution $u \equiv 0$ and

$$\alpha(f(A \times E)) \le h_{M,a}(\alpha(E))$$

for $A \subset [\min(0,a), \max(0,a)]$ and $E \subset M$.

Now we start with the following:

Theorem 4.12. *Under the above assumptions the set S of all solutions of the above* **CP**, *defined on $\mathbb{R}$ and considered as a subset of the space $C = C_b(\mathbb{R}, \mathbb{X})$ of all bounded continuous functions $\mathbb{R} \to \mathbb{X}$ with the topology of uniform convergence, is an R_δ.*

Proof. Define the mapping

$$F(x)(t) = x_0 + \int_0^t f(s, x(s))ds \quad \text{for } t \in \mathbb{R} \text{ and } x \in C.$$

Obviously $F(C) \subset C$. In view of the inequalities

$$\|F(x)(t_1) - F(x)(t_2)\| \le \int_{t_1}^{t_2} \|f(s, x(s))\|ds \le \int_{t_1}^{t_2} \phi(s)ds \quad \text{for } t_1, t_2 \in \mathbb{R},\ x \in C$$

and from the assumption about ϕ, it is clear that the family $F(C)$ is equicontinuous. Now we verify that F is a continuous operator. Let (x_n) be a sequence of elements from C such that $x_n \to x$ uniformly as $n \to \infty$. Fix $\varepsilon > 0$. First consider the halfline $\mathbb{R}_+$. Choose $t_0 > 0$ such that $2 \int_{t_0}^{+\infty} \phi(s)ds < \frac{\varepsilon}{2}$. In view of Krasnosielskii-Krein lemma there exists $\delta > 0$ such that

$$\sup_{s \in [0,t_0]} \|f(s, u(s)) - f(s, x(s))\| < \frac{\varepsilon}{2t_0},$$

whenever $u \in C$ and $\sup_{s \in [0,t_0]} \|u(s) - x(s)\| < \delta$. Choose $N \in \mathbb{N}$ such that $\sup_{s \in [0,t_0]} \|x_n(s) - x(s)\| < \delta$ for $n \geq N$. For $n \geq N$ and $t > t_0$ we have

$$\|F(x_n)(t) - F(x)(t)\| = \| \int_0^t [f(s, x_n(s)) - f(s, x(s))]ds\| \leq$$

$$\int_0^{t_0} \|f(s, x_n(s)) - f(s, x(s))\|ds + \int_{t_0}^t \|f(s, x_n(s)) - f(s, x(s))\|ds <$$

$$\frac{\varepsilon}{2} + \frac{\varepsilon}{2} = \varepsilon,$$

so $F(x_n)|_{\mathbb{R}_+} \to F(x)|_{\mathbb{R}_+}$ pointwise, as $n \to \infty$. A similar reasoning establishes that $F(x_n)|_{\mathbb{R}_-} \to F(x)|_{\mathbb{R}_-}$ pointwise, as $n \to \infty$. Since the family $F(C)$ is equicontinuous, we infer that $F(x_n) \to F(x)$ uniformly on $\mathbb{R}$, as $n \to \infty$, which proves the continuity of F.

Define

$$F_n(x)(t) = \begin{cases} x_0, & \text{if } -\frac{1}{n} \leq t \leq \frac{1}{n}, \\ x_0 + \int_0^{t-\frac{1}{n}} f(s, x(s))ds, & \text{if } t > \frac{1}{n}, \\ x_0 + \int_0^{t+\frac{1}{n}} f(s, x(s))ds, & \text{if } t < -\frac{1}{n}, \end{cases}$$

where $x \in C$ and $n \in \mathbb{N}$.

Obviously F_n maps C into itself. Essentially the same reasoning as in the case of F proves that mappings F_n are continuous. Put $T_n = I - F_n$ for $n \in \mathbb{N}$, where I denotes the identity map on C. The mappings T_n are continuous. We establish that they are homeomorphisms. Indeed, fix $n \in \mathbb{N}$. It is easy to see that for any x_1, $x_2 \in C$:

$$T_n(x_1) = T_n(x_2) \Longrightarrow x_1 = x_2.$$

Now, it is enough to prove the continuity of T_n^{-1}. Suppose (x_i) is a sequence of elements from C such that $T_n(x_i) \to T_n(x)$ uniformly as $i \to \infty$. Since

$F_n(x_i)(t) = x_0 = F_n(x)(t)$ for $|t| \le \frac{1}{n}$, $x_i \to x$ uniformly on $\{t : |t| \le \frac{1}{n}\}$, as $i \to \infty$. Further, applying the Krasnoselskii-Krein lemma, we infer that

$$\int_0^{t-\frac{1}{n}} f(s, x_i(s))ds \to \int_0^{t-\frac{1}{n}} f(s, x(s))ds \quad \text{for } \frac{1}{n} < t \le \frac{2}{n},$$

uniformly, so $x_i \to x$ on $\{t : \frac{1}{n} < t \le \frac{2}{n}\}$ uniformly, as $i \to \infty$. The similar reasoning establishes that $x_i \to x$ uniformly on $[0, t_0 + \frac{1}{n}]$, as $i \to \infty$. For $t > t_0 + \frac{1}{n}$ we have

$$\|F_n(x_i)(t) - F_n(x)(t)\| = \| \int_0^{t-\frac{1}{n}} [f(s, x_i(s)) - f(s, x(s))]ds\| \le$$

$$\int_0^{t_0} \|f(s, x_i(s)) - f(s, x(s))\|ds + \int_{t_0}^{t-\frac{1}{n}} \|f(s, x_i(s)) - f(s, x(s))\|ds$$

The above inequalities and similar arguments as in the case of the continuity of F prove that $F_n(x_i) \to F_n(x)$ uniformly on $[t_0 + \frac{1}{n}, +\infty)$, so $x_i \to x$ uniformly on this interval, as $i \to \infty$. In a similar manner we argue in the case of the halfline $\mathbb{R}_-$. Hence $x_i \to x$ uniformly on $\mathbb{R}$, as $i \to \infty$. This proves the continuity of T_n^{-1}.

Now we shall show that $\lim_{n\to\infty} T_n = T$ uniformly. We have the following inequalities

$$\|F_n(x)(t) - F(x)(t)\| = \| \int_0^t f(s, x(s))ds\| \le \int_0^{\frac{1}{n}} \phi(s)ds$$

$$\text{for } x \in C \quad \text{and } |t| \le \frac{1}{n},$$

$$\|F_n(x)(t) - F(x)(t)\| \le \int_{t-\frac{1}{n}}^t \|f(s, x(s))\|ds \le \int_{t-\frac{1}{n}}^t \phi(s)ds$$

$$\text{for } x \in C \quad \text{and } t > \frac{1}{n}.$$

In particular, for $t > t_0 + \frac{1}{n}$ we have

$$\|F_n(x)(t) - F(x)(t)\| \le \int_{t-\frac{1}{n}}^t \phi(s)ds < \frac{\varepsilon}{4}.$$

Thus $F_n(x) \to F(x)$ uniformly in x on $\mathbb{R}_+$. Arguing in a similar way we infer that $F_n(x) \to F(x)$ uniformly in x on $\mathbb{R}$.

To complete the proof it is enough to show that T is proper at $0 \in C$. Let Z be any compact set in C and put $U = T^{-1}(Z)$. Consider the sequence (u_n), where $u_n \in U$ for $n \in \mathbb{N}$. Set $V = \{u_n : n \in \mathbb{N}\}$. Since $V(t) \subset T(V)(t) + F(V)(t) \subset Z(t) + F(V)(t)$ and $Z(t)$ is compact, in view of the properties of the index α we obtain

$$\alpha(V)(t) \leq \alpha(Z(t)) + \alpha(F(V)(t)) = \alpha(F(V)(t)) \quad \text{for } t \in \mathbb{R}.$$

Thus the similar reasoning as in the proof of Theorem 3 [Bugajewska (16)] proves that $V(t)$ is compact for every $t \in \mathbb{R}$. In view of Ascoli's theorem we infer that V is relatively compact and thus (u_n) has a limit point. Therefore U is a compact, so T is proper. In view of *Theorem 6.7*, $T^{-1}(0)$ is an R_δ set. Since $T^{-1}(0)$ coincides with the set of all fixed points of F, i.e. it coincides with S, the proof is complete. □

Now assume

3^0 $(s, x) \to f(s, x)$ is a function from $\mathbb{R} \times \mathbb{X}$ to $\mathbb{X}$, which is almost automorphic in s for each x and satisfies a Lipschitz condition in x uniformly in $s \in \mathbb{R}$;

4^0 for each $x \in AA(\mathbb{X})$ the range $R_{F(x)} = \{F(x)(s) : s \in \mathbb{R}\}$, where F is the operator defined in the proof of the above theorem, is relatively compact.

We now give the following important result:

Theorem 4.13. *Suppose* 2^0, 3^0, 4^0 *are satisfied and that the function* f *satisfies the inequality as in* 1^0. *Then* **CP** *has a solution in* $AA(\mathbb{X})$. *Moreover, the set* S *of all such solutions is compact.*

Proof. In view of 3^0 and [N'Guérékata (79)] Theorem 2.2.6, the function $s \to f(s, x(s))$, $s \in \mathbb{R}$ is almost automorphic for any $x \in AA(\mathbb{X})$. By 4^0 and [N'Guérékata (79)], Theorem 2.4.4, $F(x) \in AA(\mathbb{X})$ for any $x \in AA(\mathbb{X})$. Hence F maps $AA(\mathbb{X})$ into itself. Let $D = \overline{\text{conv}}(F(AA(\mathbb{X})))$. It is clear that F maps D into itself. Essentially the same reasoning as in the proof of Theorem 3 proves that the family $F(AA(\mathbb{X}))$ is equicontinuous and that F is a continuous mapping. Now, let V be a subset of D such that $V \subset$

$\overline{\text{conv}}(F(V) \cup \{0\})$, where $0 \in AA(\mathbb{X})$. In view of the properties of the index α we obtain

$$\alpha(V(t)) \le \alpha(\overline{\text{conv}}F(V)(t) \cup \{0\}) = \alpha(F(V)(t) \cup \{0\}) =$$
$$\max(\alpha(F(V)(t)), \alpha(\{0\})) = \alpha(F(V)(t)) \qquad \text{for } t \in \mathbb{R},$$

and therefore the same reasoning as in the proof of Theorem 3 [Bugajewska (16)] proves that $V(t)$ is relatively compact for $t \in \mathbb{R}$. In view of Ascoli's theorem we infer that V is relatively compact. The operator F satisfies all assumptions of Theorem 1 and, therefore, there exists $z \in D$ such that $z = F(z)$.

Finally, since $S = F(S)$, the same reasoning as in Theorem 3 [Bugajewska (16)] proves that S is relatively compact. Since S is closed, it is compact, which completes the proof. □

Remark 4.7. In the case of uniformly convex Banach spaces the assumption 4^0 in Theorem 4 can be weaken. Namely, in view of [N'Guérékata (79)], Theorem 2.4.6 it is enough to assume then that the range $R_{F(x)}$ is bounded for each $x \in AA(\mathbb{X})$.

Remark 4.8. By the assumption 3^0 and [N'Guérékata (79)] Theorem 2.2.6, the function $s \to f(s, x(s))$, $s \in \mathbb{R}$ is almost automorphic for each $x \in AA(\mathbb{X})$. Thus by [N'Guérékata (79)] Theorem 2.2.6 the range $\{f(s, x(s)) : s \in \mathbb{R}\}$ is relatively compact in $\mathbb{X}$ for each $x \in AA(\mathbb{X})$. Further, since $F(x)$ is almost automorphic if and only if the range $R_{F(x)}$ is relatively compact in $\mathbb{X}$ ([N'Guérékata (79)], Theorem 2.4.4.), the assumption 4^0 is natural in the considered situation.

4.7 Comments and Further Reading Guide

Recently, numerous contributions were made to the study of abstract evolution equations with almost automorphic solutions. These include work by D. Bugajewski, T. Diagana, S. G. Gal, S. C. Gal, J. A. Goldstein, K. Ezzinbi, Y. Hino, J. Liu, S. Murakami, G. M. N'Guérékata, Nguyen Van Minh, A. S. Rao, S. Zaidman, etc.

Note that Theorem 4.2 is usually referred as Bohr-Neugebauer type theorem. The proof presented here is a slight modification of Theroem 4.2.2 [N'Guérékata (79)]. There exist several generalizations of the Bohr-Neuhegauer Theorem. We send the reader for instance to: [N'Guérékata (81)], [N'Guérékata (83)], [N'Guérékata (88)].

Existence of almost automorphic solutions to differential equations is a fundamental problem still under investigation. Various methods were used including fixed point theorems for linear equations $x' = Ax(t) + f(t)$ ([N'Guérékata (87)]), as well as nonlinear and inte-grodifferential equations ([Bugajewski and N'Gurkata (17)]) and semilinear equation $x' = Ax(t) + f(t,x)$ ([N'Guérékata (79)], [N'Guérékata (88)]) when A generates an asymptotically stable semigroup of linear operators; the method of invariant subspaces, introduced in 2003 by T. Diagana and G. M. N'Guérékata for perturbed equations of the form $x' = (A+B)x + f(t)$ where both operators A and B are unbounded; the method of sums of commuting operators (Theorems 4.9, 4.11). In this latter case, the challenge is to overcome the difficulty due to the fact that almost automorphic functions are not necessarily uniformly continuous. This implies the non-strong continuity of translation semigroup in the function space $AA(X)$. This approach, as well as the method of invariant subspaces are also used successfully in the case of higher order differential equations in various recent papers by T. Diagana, J. Liu, G. M. N'Guérékata, and Nguyen van Minh.

Readers interested in fuzzy settings of the theory of almost automorphy may consult [N'Guérékata (88)] and other papers by S. G. Gal and G. M. N'Guérékata.

Chapter 5

Nonlinear equations

5.1 Periodic Solutions of Nonlinear equations

5.1.1 *Nonlinear Equations Without Delay*

In this section, we study the existence of periodic solutions of the following nonlinear differential equation without delay in a general Banach space $\mathbb{X} = (\mathbb{X}, \|\cdot\|)$,

$$u'(t) + A(t)u(t) = f(t, u(t)), \quad t > 0, \quad u(0) = u_0. \tag{5.1}$$

To this end, we make the following assumptions.

Assumption 5.1. For a constant $T > 0$, $A(t+T) = A(t)$, $t \geq 0$. The function f is continuous in all its variables, T-periodic in the first variable t and Lipschitzian in the other variables uniformly in t.

Remark 5.1. We used "variable(s)" in the assumption 5.1 because then it can be used for other cases where the function f is of more than two variables.

Assumption 5.2. For $t \in [0, T]$,

(H1). The domain $D(A(t)) = D$ is independent of t and is dense in $\mathbb{X}$.

(H2). For $t \geq 0$, the resolvent $R(\lambda, A(t)) = (\lambda I - A(t))^{-1}$ exists for all λ with $Re\lambda \leq 0$ and is compact, and there is a constant M independent of λ and t such that

$$\|R(\lambda, A(t))\| \leq M(|\lambda| + 1)^{-1}, \quad Re\lambda \leq 0.$$

(H3). There exist constants L and $0 < a \leq 1$ such that

$$\|(A(t) - A(s))A(r)^{-1}\| \leq L|t - s|^a, \quad s, t, r \in [0, T].$$

Under these assumptions, the results in, e.g., Pazy [Pazy (90)], imply the existence of a unique evolution system $U(t,s)\,(0 \le s \le t \le T)$ for Eq. (5.1). The materials presented here are based on [Liu (60)].

Theorem 5.1. *Let the assumptions 5.1 and 5.2 be satisfied and let $u_0 \in \mathbb{X}$. Then there exists a constant $\alpha > 0$ such that Eq. (5.1) has a unique mild solution (also called solution here) $u : [0,\alpha] \to \mathbb{X}$ satisfying*

$$u(t) = U(t,0)u_0 + \int_0^t U(t,h)f(h,u(h))dh, \quad t \in [0,\alpha]. \tag{5.2}$$

Proof. We only need to set up the framework for the use of the contraction mapping principle. With $u_0 \in \mathbb{X}$ being fixed and with $\alpha > 0$ yet to be determined, we define a map (operator) Q on $C([0,\alpha],\mathbb{X})$ such that

$$(Qu)(t) = U(t,0)u_0 + \int_0^t U(t,h)f(h,u(h))dh, \quad \text{for} \quad t \in [0,\alpha]. \tag{5.3}$$

Using the property of the evolution system U, we have $Q : C([0,\alpha],\mathbb{X}) \to C([0,\alpha],\mathbb{X})$. Next, for $u,\ v \in C([0,\alpha],\mathbb{X})$, one has

$$(Qu)(t) - (Qv)(t) = \int_0^t U(t,h)\Big[f(h,u(h)) - f(h,v(h))\Big]dh. \tag{5.4}$$

Now, f is Lipschitzian in the second variable and $U(t,h)$ is a bounded operator, it is then clear that we can obtain the result by using the contraction mapping principle. Details are left in an exercise. □

Note that we are concerned with periodic solutions here, so we assume that solutions exist on $[0,\infty)$. We will write $u = u(\cdot,u_0)$ for the unique solution with the initial value u_0.

Next, we give some basic results concerning the search of periodic solutions.

Lemma 5.1. *Let the assumptions 5.1 and 5.2 be satisfied.*

(a). If $u(t)$ is a solution of Eq. (5.1), then so is $u(t+T)$, $t \ge 0$.
(b). Let $u(t,u_0)$ be a solution of Eq. (5.1) with $u(0,u_0) = u_0$. Then $u(t,u_0)$ is T-periodic if and only if $u(T,u_0) = u_0$.

Proof. (a): Let $y(t) = u(t+T)$. Now, for $t \ge 0$, we can use the known formulas ([Pazy (90)]) $U(t,s) = U(t,r)U(r,s),\ 0 \le s \le r \le t \le T$, and $U(t+T, s+T) = U(t,s)$ (since the operator $A(t)$ is T-periodic in t) to obtain:

$$
\begin{aligned}
y(t) = u(t+T) &= U(t+T,0)u_0 + \int_0^{t+T} U(t+T,h)f(h,u(h))dh \\
&= U(t+T,T)U(T,0)u_0 + \int_0^{T} U(t+T,h)f(h,u(h))dh \\
&\quad + \int_T^{t+T} U(t+T,h)f(h,u(h))dh \\
&= U(t,0)U(T,0)u_0 + \int_0^{T} U(t+T,T)U(T,h)f(h,u(h))dh \\
&\quad + \int_0^{t} U(t+T,T+s)f(T+s,u(T+s))ds \\
&= U(t,0)U(T,0)u_0 + \int_0^{T} U(t,0)U(T,h)f(h,u(h))dh \\
&\quad + \int_0^{t} U(t,s)f(s,y(s))ds \\
&= U(t,0)\Big[U(T,0)u_0 + \int_0^{T} U(T,h)f(h,u(h))dh\Big] \\
&\quad + \int_0^{t} U(t,s)f(s,y(s))ds \\
&= U(t,0)u(T) + \int_0^{t} U(t,s)f(s,y(s))ds \\
&= U(t,0)y(0) + \int_0^{t} U(t,s)f(s,y(s))ds. \qquad (5.5)
\end{aligned}
$$

The equality (5.5) implies that $y(t) = u(t+T)$ is also a solution of Eq. (5.1).

(b): If $u(t,u_0)$ is a T-periodic solution of Eq. (5.1), then $u(T,u_0) = u(0,u_0) = u_0$. On the other hand, if $u(T,u_0) = u_0$, then, from (a), $y(t) = u(t+T,u_0)$ is also a solution of Eq. (5.1) with $y(0) = u(T,u_0) = u_0$. By uniqueness, $y(t) = u(t, u_0)$, or $u(t+T, u_0) = u(t, u_0)$, $t \geq 0$, thus $u(t, u_0)$ is T-periodic. This completes the proof. □

Accordingly, we can define a map $P : \mathbb{X} \to \mathbb{X}$ (called a *Poincaré map* or a *Poincaré operator*) such that for each $u_0 \in \mathbb{X}$ and for the corresponding unique solution $u(t) = u(t, u_0)$ with $u(0, u_0) = u_0$, define

$$P(u_0) = u(T) = U(T,0)u_0 + \int_0^T U(T,r)f(r, u(r, u_0))dr. \tag{5.6}$$

Now, Lemma 5.1 (b) indicates that Eq. (5.1) has a T-periodic solution if and only if there exists an $u_0 \in \mathbb{X}$ such that $P(u_0) = u_0$. This is the same as saying that the Poincaré map P has a **fixed point**. We formulate this as follows.

Lemma 5.2. *Eq. (5.1) has a T-periodic solution if and only if the Poincaré map $P : \mathbb{X} \to \mathbb{X}$ defined in (5.6) has a fixed point.*

We will see that Lemma 5.2 provides a very useful approach for deriving periodic solutions, as it reduces the search of periodic solutions to that of fixed points of the Poincaré map P, for which some well-known fixed point theorems from Functional Analysis can be applied.

By Massera's theorems for nonlinear differential equations in $\Re^n$, $n \leq 2$ ([Liu (59)]), the existence of *one bounded solution* is good enough to guarantee the existence of a periodic solution. However, for nonlinear differential equations in $\Re^n$, $n \geq 3$, the situations become very complicated. For example, for the Lorenz equation in $\Re^3$ ([Liu (59)]) where chaos occurs, the solutions may be bounded, but periodic solutions may not exist.

Next, note that if u_0 is in some set, then $P(u_0) = u(T, u_0)$ may not be in the same set. Therefore, we will make use of some kind of *asymptotic fixed point theorems*, where the idea is to let P^k (for some positive integer k) map a set into the same set and then prove the existence of a unique fixed point of P^k, which then implies the existence of a fixed point of P (not necessarily unique). Among the asymptotic fixed point theorems, Horn's fixed point theorem is easy to use with the notion of boundedness of solutions. So we state Horn's fixed point theorem here and make the following definitions concerning boundedness of solutions.

Theorem 5.2. *(Horn's fixed point theorem) Let $E_0 \subset E_1 \subset E_2$ be convex subsets of a Banach space Z, with E_0 and E_2 compact subsets and E_1 open relative to E_2. Let $P : E_2 \to Z$ be a continuous operator such that for some* *integer m, one has*

$$P^j(E_1) \subset E_2, \quad 1 \le j \le m-1, \tag{5.7}$$

$$P^j(E_1) \subset E_0, \quad m \le j \le 2m-1, \tag{5.8}$$

then P has a fixed point in E_2.

Definition 5.1. [Burton (18)] The solutions of Eq. (5.1) are said to be **bounded** if for each $B_1 > 0$ there is a $B_2 = B_2(B_1) > 0$ such that $\{\|u_0\| \le B_1, t \ge 0\}$ implies $\|u(t, u_0)\| < B_2$.

Definition 5.2. [Burton (18)] The solutions of Eq. (5.1) are said to be **ultimate bounded** if there is a bound $B > 0$ such that for each $B_3 > 0$, there is a $K = K(B, B_3) > 0$ such that $\{\|u_0\| \le B_3, t \ge K\}$ implies $\|u(t, u_0)\| < B$.

Definition 5.3. An operator $S : \mathbb{X} \to \mathbb{X}$ is said to be a **compact operator** on $\mathbb{X}$ if S is continuous and maps a bounded set into a precompact set.

For the relationship between boundedness and ultimate boundedness, we have the following result.

Theorem 5.3. *Assume that $f(t, u)$ is continuous and is Lipschitzian in u. If the solutions of Eq. (5.1) are ultimately bounded, then they are also bounded.*

Proof. Let $B > 0$ be the bound in the definition of ultimate boundedness, then for any $B_1 > 0$, there is a $K = K(B, B_1) > 0$ such that $\{\|u_0\| \le B_1,\ t \ge K\}$ imply $\|u(t, u_0)\| \le B$. Next, for t in the interval $[0, K]$, we have

$$\begin{aligned}
\|u(t, u_0)\| &\le \|U(t, 0)u_0\| + \int_0^t \|U(t, s)f(s, u(s, u_0))\| ds \\
&\le \|U(t, 0)u_0\| + \int_0^t \|U(t, s)[f(s, u(s, u_0)) - f(s, 0)]\| ds \\
&\quad + \int_0^t \|U(t, s)f(s, 0)\| ds \\
&\le \|U(t, 0)u_0\| + c\int_0^t |U(t, s)| \|u(s, u_0)\| ds \\
&\quad + \int_0^t \|U(t, s)f(s, 0)\| ds,
\end{aligned}$$

where c is a Lipschitz constant. For $0 \leq s \leq t \leq K$, there are positive constants G_1 and G_2 such that

$$|U(t,s)| \leq G_1, \quad |U(t,0)|B_1 + \int_0^t \|U(t,s)f(s,0)\|ds \leq G_2.$$

Then for $t \in [0,K]$ and $\|u_0\| \leq B_1$,

$$\|u(t,u_0)\| \leq G_2 + c\int_0^t G_1\|u(s,u_0)\|ds.$$

Hence, Gronwall's inequality implies, for $t \in [0,K]$ and $\|u_0\| \leq B_1$,

$$\|u(t,u_0)\| \leq G_2 e^{cG_1 t} \leq G_2 e^{cG_1 K}, \tag{5.9}$$

which implies that for $\|u_0\| \leq B_1$, the solutions $u(t,u_0)$ are bounded on $[0,K]$. That is, there is a $B_3 = B_3(B_1,K) = B_3(B_1,B) > 0$ such that $\|u(t,u_0)\| \leq B_3$ for $t \in [0,K]$ and $\|u_0\| \leq B_1$. Now, we see that $\{\|u_0\| \leq B_1,\, t \geq 0\}$ imply $\|u(t,u_0)\| \leq B_2 = \max\{B_3, B\}$, which verifies the boundedness of the solutions. This completes the proof. $\square$

To derive periodic solutions, the following lemma from Amann [Amann (2)] will be used to show that the Poincaré map P defined in (5.6) is a compact operator. Recall that in the usual way (see, e.g., Amann [Amann (2)], Pazy [Pazy (90)]) we define fractional power operator A^α and Banach space $\mathbb{X}_\alpha$ for $0 \leq \alpha \leq 1$, where $A = A(0)$ and $\mathbb{X}_\alpha = (D(A^\alpha), \|\cdot\|_\alpha)$ with $\|x\|_\alpha \equiv \|A^\alpha x\|$. We also write the norm in $L(\mathbb{X}_\alpha, \mathbb{X}_\beta)$ (space of bounded linear operators from $\mathbb{X}_\alpha$ to $\mathbb{X}_\beta$) as $\|\cdot\|_{\alpha,\beta}$.

Lemma 5.3. *[Amann (2)] (i) Suppose that $0 \leq \alpha \leq \beta < 1$. Then for $\beta - \alpha < \gamma < 1$, there is a constant $C(\alpha,\beta,\gamma)$ such that*

$$\|U(t,h)\|_{\alpha,\beta} \leq C(\alpha,\beta,\gamma)(t-h)^{-\gamma}, \quad 0 \leq h < t \leq T.$$

(ii) For $0 \leq \gamma < 1$, there is a constant $C(\gamma)$, such that for $g \in C([0,L],\mathbb{X})$ ($L > 0$ is a constant), one has for $0 \leq s,t \leq L$,

$$\|\int_0^t U(t,h)g(h)dh - \int_0^s U(s,h)g(h)dh\| \leq C(\gamma)|t-s|^\gamma \max_{0 \leq h \leq L} \|g(h)\|.$$

(iii) Let $0 \leq \alpha < \beta \leq 1$. Then

$$K(x,g)(t) \equiv U(t,0)x + \int_0^t U(t,h)g(h)dh, \quad 0 \leq t \leq T,$$

defines a continuous linear operator from $\mathbb{X}_\beta \times C([0,T],\mathbb{X})$ into $C^\gamma([0,T],\mathbb{X}_\alpha)$ for every $\gamma \in [0,\beta-\alpha)$.

Now we prove that the Poincaré map $P : \mathbb{X} \to \mathbb{X}$ defined in (5.6) is compact.

Theorem 5.4. *Let assumptions 5.1 and 5.2 be satisfied and assume that solutions of Eq. (5.1) are bounded. Then the Poincaré map $P : \mathbb{X} \to \mathbb{X}$ defined in (5.6) is a compact operator.*

Proof. The continuity of P is left as an exercise. Next, fix $\eta \in (0,1)$. Then from Lemma 5.3 (i), $U(T,0) : \mathbb{X} \to \mathbb{X}_\eta$ is bounded. Next, let E be bounded in $\mathbb{X}$. Since solutions of Eq. (5.1) are bounded, there exists $M_1 = M_1(E,f) > 0$ such that

$$\|f(t,u(t,u_0))\| \leq M_1, \quad t \in [0,T], \quad u_0 \in E.$$

Now for $u_0 \in E$, we let $g(t) = f(t,u(t,u_0))$. Then

$$\int_0^T U(T,r)f(r,u(r,u_0))dr \equiv K(0,g)(T) \in \mathbb{X}_\eta$$

according to Lemma 5.3 (ii). Also note that by Lemma 5.3 (i), there are constants $\gamma \in (0,1)$ and $M_2 > 0$ such that

$$\|U(T,s)\|_{0,\eta} \leq M_2(T-s)^{-\gamma}, \quad 0 \leq s < T.$$

Thus

$$\|\int_0^T U(T,r)f(r,u(r,u_0))dr\|_\eta \leq M_1M_2T^{1-\gamma}/(1-\gamma), \quad u_0 \in E.$$

Therefore $P : \mathbb{X} \to \mathbb{X}_\eta$ is bounded. Next, the embedding $\mathbb{X}_\eta \to \mathbb{X}$ is compact (see, e.g., [Hale (40)]), thus $P : \mathbb{X} \to \mathbb{X}$ is a compact operator. This completes the proof. □

Now, we are ready to state and prove the existence of periodic solutions for Eq. (5.1) without delay.

Theorem 5.5. *Let the assumptions 5.1 and 5.2 be satisfied. If the solutions of Eq. (5.1) are ultimate bounded, then Eq. (5.1) has a T-periodic solution.*

Proof. From Theorem 5.3, the solutions of Eq. (5.1) are also bounded. Let B be the bound in the definition of ultimate boundedness, then by boundedness, there is a $B_1 > 0$ such that $\|u_0\| \leq B$ implies $\|u(t,u_0)\| < B_1$ for $t \geq 0$. Furthermore, there is a $B_2 > B_1$ such that $\|u_0\| \leq B_1$ implies $\|u(t,u_0)\| < B_2$ for $t \geq 0$. Now, using ultimate boundedness, there is a positive integer m such that $\|u_0\| \leq B_1$ implies $\|u(t,u_0)\| < B$ for $t \geq (m-1)T$.

We know from Lemma 5.1 that $u(T+t, u_0)$ is also a solution, then the uniqueness implies that $u(T+t, u_0) = u(t, u(T, u_0))$, $t \geq 0$. Letting $t = T$, this means $P^2(u_0) = u(2T, u_0)$. By an induction, it can be shown (see an exercise) that

$$P^k(u_0) = u(kT, u_0), \; k \geq 1. \tag{5.10}$$

Thus, we have the following

$$\|P^{j-1}(u_0)\| = \|u((j-1)T, u_0)\| < B_2, \;\; j = 1, 2, ..., m-1 \;\text{ and }\; \|u_0\| \leq B_1,$$

$$\|P^{j-1}(u_0)\| = \|u((j-1)T, u_0)\| < B, \;\; j \geq m \;\text{ and }\; \|u_0\| \leq B_1.$$

Now let

$$\begin{cases} H = \{u_0 \in \mathbb{X} : \|u_0\| < B_2\}, \; E_2 = \overline{cov.(P(H))}, \\ K = \{u_0 \in \mathbb{X} : \|u_0\| < B_1\}, \; E_1 = K \cap E_2, \\ G = \{u_0 \in \mathbb{X} : \|u_0\| < B\}, \;\; E_0 = \overline{cov.(P(G))}, \end{cases} \tag{5.11}$$

where $cov.(F)$ is the *convex hull* of the set F defined by $cov.(F) = \{\sum_{i=1}^{n} \lambda_i f_i : n \geq 1, f_i \in F, \lambda_i \geq 0, \sum_{i=1}^{n} \lambda_i = 1\}$. From Theorem 5.4, the operator P is compact. It is also known that a convex hull of a precompact set is also precompact, then we see that E_0, E_1 and E_2 are convex subsets of $\mathbb{X}$ with E_0 and E_2 compact subsets and E_1 open relative to E_2. Next, from (5.11), we have

$$P^j(E_1) \subset P^j(K) = PP^{j-1}(K) \subset P(H) \subset E_2, \;\; 1 \leq j \leq m-1,$$

$$P^j(E_1) \subset P^j(K) = PP^{j-1}(K) \subset P(G) \subset E_0, \;\; m \leq j \leq 2m-1.$$

Consequently, we know from Horn's fixed point theorem that the operator P has a fixed point $u_0 \in \mathbb{X}$. This completes the proof by using Lemma 5.2.

□

Next, we provide a result which asserts that the existence of a proper Liapunov function implies boundedness and ultimate boundedness of solutions.

Theorem 5.6. *Assume that there exist functions ("wedges") $W_i, i = 1, 2, 3$, with $W_i : [0, \infty) \to [0, \infty), W_i(0) = 0, W_i$ strictly increasing, and $W_1(t) \to \infty, t \to \infty$. Further, assume that there exists a (Liapunov) function $V : \mathbb{X} \to \Re$ (reals) such that for some constant $M > 0$, when u is a solution of Eq. (5.1) with $\|u(t)\| \geq M$, then*

(a). $W_1(\|u(t)\|) \leq V(u(t)) \leq W_2(\|u(t)\|)$, and

(b). $\frac{d}{dt}V(u(t)) \le -W_3(\|u(t)\|)$.

Then solutions of Eq. (5.1) are bounded and ultimate bounded.

Proof. Let $u(t) = u(t, u_0)$ and let $B_1 > 0$ be given with $B_1 \ge M$. Find $B_2 \ge B_1$ with $W_1(B_2) = W_2(B_1)$. If for some interval $[t_1, t_2]$ with $t_1 \ge 0, \|u(t_1)\| = B_1$, and $\|u(t)\| \ge B_1$ on $[t_1, t_2]$, then for $t \in [t_1, t_2]$,

$$\begin{aligned} W_1(\|u(t)\|) &\le V(u(t)) \le V(u(t_1)) \\ &\le W_2(\|u(t_1)\|) = W_2(B_1) = W_1(B_2), \end{aligned} \tag{5.12}$$

and hence $\|u(t)\| \le B_2$ for $t \in [t_1, t_2]$. This implies that if $\|u_0\| \le B_1$, then $\|u(t)\| \le B_2, t \ge 0$, which gives the boundedness.

Next, find $B \ge M + 1$ with $W_1(B) = W_2(M + 1)$. Then, similar to (5.12), one shows, for $t_1 \ge 0$,

$$\begin{aligned} &\Big\{\|u(t_1)\| = M + 1, \|u(t)\| \ge M + 1, t \in [t_1, t_2]\Big\} \\ &\qquad \Rightarrow \Big\{\|u(t)\| \le B, t \in [t_1, t_2]\Big\}. \end{aligned} \tag{5.13}$$

Let $B_3 > 0$ be given. We need to prove that there is a $K > 0$ such that if $\|u_0\| \le B_3$ and $t \ge K$, then $\|u(t)\| \le B$.

According to (5.13), we only need to show that there is a $K > 0$ such that when $\|u_0\| \le B_3$, there is a $t^0 \in [0, K]$ with $\|u(t^0)\| \le M+1$. (Because then if there is a $t_2 \ge K$ with $\|u(t_2)\| > B \ge M + 1$, we can find such a $t_1 < t_2$ that $\|u(t_1)\| = M+1$ and that (5.13) can be used to get $\|u(t_2)\| \le B$, a controdiction.)

Now, if $\|u(t)\| > M + 1$ for $t \ge 0$, then

$$0 < W_1(M+1) \le V(u(t)) \le V(u_0) - W_3(M+1)t \le W_2(B_3) - W_3(M+1)t.$$

This fails when $t \ge W_2(B_3)/W_3(M + 1)$. Thus we can choose $K \equiv W_2(B_3)/W_3(M + 1)$ to finish the proof. □

Exercise 20.

(1) Complete the proof of Theorem 5.1.
(2) In the proof of Theorem 5.4, show that the Poincaré map $P : \mathbb{X} \to \mathbb{X}$ defined in (5.6) is continuous.
(3) In the proof of Theorem 5.5, use an induction to show that

$$P^k(u_0) = u(kT, u_0), \ k \ge 1.$$

(4) Prove that a convex hull of a precompact set is also precompact.

5.1.2 *Nonlinear Equations With Finite Delay*

Consider the following nonlinear differential equation with finite delay,

$$u'(t) + A(t)u(t) = f(t, u(t), u_t), \;\; t > 0, \;\; u(s) = \phi(s), \;\; s \in [-r, 0], \tag{5.14}$$

in a general Banach space $\mathbb{X}$, where $r > 0$ is a constant, and $u_t(s) = u(t+s), s \in [-r, 0]$.

The materials presented here are based on [Liu (61)]. We still assume that the assumptions 5.1 and 5.2 are satisfied, so similar to the previous section, we have

Theorem 5.7. *Let the assumptions 5.1 and 5.2 be satisfied and let $\phi \in C([-r, 0], \mathbb{X})$ (the Banach space with the sup-norm). Then there exists a constant $\alpha > 0$ such that Eq. (5.14) has a unique (mild) solution $u : [-r, \alpha] \to \mathbb{X}$ satisfying $u_0 = \phi$ (i.e., $u(s) = \phi(s)$, $s \in [-r, 0]$), and*

$$u(t) = U(t, 0)\phi(0) + \int_0^t U(t, h)f(h, u(h), u_h)dh, \;\; t \in [0, \alpha]. \tag{5.15}$$

To derive periodic solutions, we assume that solutions exist on $[0, \infty)$. We will write $u = u(\cdot, \phi)$ for the unique solution with the initial function ϕ.

Note that for equations without delay, the Poincaré operator maps a single element of the Banach space $\mathbb{X}$ to a single element of $\mathbb{X}$. Now, for equations with finite delay, the difference is that the Poincaré operator will map an element ϕ of $C([-r, 0], \mathbb{X})$ (thus a *function*) to an element of $C([-r, 0], \mathbb{X})$, given by

$$P\phi = u_T(\cdot, \phi), \;\; \phi \in C([-r, 0], \mathbb{X}), \tag{5.16}$$

or, for $s \in [-r, 0]$,

$$(P\phi)(s) = u_T(s, \phi) = u(T + s, \phi)$$

$$= \begin{cases} U(T+s, 0)\phi(0) + \displaystyle\int_0^{T+s} U(T+s, h)f(h, u(h), u_h)dh, \\ \qquad\qquad\qquad\qquad\qquad\qquad\qquad\qquad T + s > 0, \\ \phi(T+s), \;\; T + s \le 0. \end{cases}$$

Similar to lemmas 5.1 and 5.2, we have

Lemma 5.4. *Let the assumptions 5.1 and 5.2 be satisfied.*

(a). If $u(t)$ is a solution of Eq. (5.14), then so is $u(t + T)$, $t \ge 0$.

(b). Equation (5.14) has a T-periodic solution if and only if the Poincaré map $P : C([-r,0],\mathbb{X}) \to C([-r,0],\mathbb{X})$ defined in (5.16) has a fixed point.

Definitions concerning boundedness and ultimate boundedness given in the previous section can be revised as follows to suit Eq. (5.14).

Definition 5.4. Denote $C = C([-r,0],\mathbb{X})$. The solutions of Eq. (5.14) are said to be **bounded** if for each $B_1 > 0$ there is a $B_2 = B_2(B_1) > 0$ such that $\{|\phi|_C \leq B_1, t \geq 0\}$ implies $\|u(t,\phi)\| < B_2$. (Here $|\cdot|_C$ means the sup-norm.)

Definition 5.5. The solutions of Eq. (5.14) are said to be **ultimate bounded** if there is a bound $B > 0$ such that for each $B_3 > 0$, there is a $K = K(B, B_3) > 0$ such that $\{|\phi|_C \leq B_3, t \geq K\}$ implies $\|u(t,\phi)\| < B$.

For the relationship between boundedness and ultimate boundedness, we also have the following result.

Theorem 5.8. *Assume that $f(t,u,v)$ in Eq. (5.14) is continuous and is Lipschitzian in u and in v. If the solutions of Eq. (5.14) are ultimately bounded, then they are also bounded.*

From the study for equations without delay, we see that for Eq. (5.14) with finite delay, the most important step will be the proof of the compactness of the Poincaré operator defined in (5.16). This is getting a little hard since now we need to deal with functions rather than single elements. The following Ascoli-Arzela theorem for general Banach spaces is needed here to treat a set of functions.

Theorem 5.9. *(Ascoli-Arzela) Let $E \subset C([-r,0],\mathbb{X})$ be bounded. Then E is precompact if and only if functions in E are equicontinuous and for each $t \in [-r,0]$, the set $\{f(t) : f \in E\}$ is precompact in $\mathbb{X}$.*

Accordingly, for a bounded set $E \subset C([-r,0],\mathbb{X})$, we need to prove the precompactness of

$$[P(E)](s) = \{(P\phi)(s) : \phi \in E\}$$

for every $s \in [-r,0]$.

Now, note from geometry that the operator P defined in (5.16) maps an initial function ϕ on $[-r,0]$ to the function $P\phi$ on $[T-r,T]$ along the unique solution $u(\cdot,\phi)$. If $T-r \leq 0$, then the restrictions of $P\phi$ on $[T-r,0]$ for $\phi \in E$ are parts of the initial functions $\phi \in E$, and they may be arbitrary, or "bad", i.e., noncompact. Therefore, to avoid this, we require $T - r > 0$. Now, for $\phi \in C([-r,0],\mathbb{X})$, $P\phi = u_T(\cdot,\phi)$ is a function defined on $[T-r,T] \subset (0,\infty)$, thus the possibly "bad" history on $[-r,0]$ is cut off. Then the operator P can smooth things out so that the same idea for equations without delay and Horn's fixed point theorem can be applied to derive periodic solutions.

Theorem 5.10. *Let the assumptions 5.1 and 5.2 be satisfied and let $T > r$. If the solutions of Eq. (5.14) are bounded, then $P : C([-r,0],\mathbb{X}) \to C([-r,0],\mathbb{X})$ defined in (5.16) is a compact operator.*

Proof. The continuity of P is left as an exercise. Let $H \subset C([-r,0],\mathbb{X})$ be bounded. Since the solutions of Eq. (5.14) are bounded, it follows that $E = P(H) \subset C([-r,0],\mathbb{X})$ is bounded. In the following, we will use the Ascoli-Arzela theorem to show that E is precompact.

Note that $T - r > 0$, so for $s \in [-r,0]$, a function in E can be expressed as

$$\begin{aligned}(P\phi)(s) &= u_T(s,\phi) = u(T+s,\phi)\\ &= U(T+s,0)\phi(0) + \int_0^{T+s} U(T+s,h)f(h,u(h),u_h)dh,\ \phi \in H.\end{aligned}$$

Also, as $T - r > 0$, there is a $k > 0$ such that $T + s > k$ for $s \in [-r,0]$. From the properties for the evolution system $U(t,s)$, one has, for $s \in [-r,0]$,

$$U(T+s,0)\phi(0) = U(T+s,k)U(k,0)\phi(0), \quad \phi \in H. \tag{5.17}$$

Fix $\eta \in (0,1)$. Then from Lemma 5.3 (i), $U(k,0) : \mathbb{X} \to \mathbb{X}_\eta$ is bounded. Next the embedding $\mathbb{X}_\eta \to \mathbb{X}$ is compact, thus $\{U(k,0)\phi(0) : \phi \in H\}$ is precompact in $\mathbb{X}$ since $\{\phi(0) : \phi \in H\}$ is bounded in $\mathbb{X}$. Therefore, the closure of $\{U(k,0)\phi(0) : \phi \in H\}$ is compact in $\mathbb{X}$. Now, one can verify that as functions on $\cdot \in [-r,0]$,

$$\{U(T+\cdot,0)\phi(0) : \phi \in H\} = \{U(T+\cdot,k)\Big[U(k,0)\phi(0)\Big] : \phi \in H\} \tag{5.18}$$

is equicontinuous. Next, from Lemma 5.3 (ii), for $0 \le \gamma < 1$, there is a constant $C(\gamma)$, such that for s_1, $s_2 \in [-r,0]$,

$$\begin{aligned}&\|\int_0^{T+s_2} U(T+s_2,h)f(h,u(h),u_h)dh - \int_0^{T+s_1} U(T+s_1,h)f(h,u(h),u_h)dh\|\\ &\qquad \le C(\gamma)|s_1 - s_2|^\gamma \max_{0\le h\le T} \|f(h,u(h),u_h)\|.\end{aligned}$$

Since solutions of Eq. (5.14) are bounded and f maps a bounded set into a bounded set, there exists $M_1 = M_1(H) > 0$ such that

$$\|f(t, u(t, \phi), u_t(\phi))\| \leq M_1, \quad t \in [0, T], \quad \phi \in H.$$

Thus, as functions on $\cdot \in [-r, 0]$,

$$\{\int_0^{T+\cdot} U(T + \cdot, h) f(h, u(h), u_h) dh : \phi \in H\} \tag{5.19}$$

is also equicontinuous. Therefore, functions in E are equicontinuous.

Next, to check the precompactness of the functions at every point of $[-r, 0]$, we fix $s_0 \in [-r, 0]$. From the above arguments, we also know that

$$\{U(T + s_0, 0)\phi(0) : \phi \in H\} \tag{5.20}$$

is precompact in $\mathbb{X}$. Now for $\phi \in H$, we let $g(t) = f(t, u(t, \phi), u_t(\phi))$. Then

$$\int_0^{T+s_0} U(T + s_0, h) f(h, u(h, \phi), u_h(\phi)) dh \equiv K(0, g)(T + s_0) \in \mathbb{X}_\eta$$

according to Lemma 5.3 (iii). Also note that by Lemma 5.3 (i), there are constants $\gamma \in (0, 1)$ and $M_2 > 0$ such that

$$\|U(T + s_0, h)\|_{0,\eta} \leq M_2(T + s_0 - h)^{-\gamma}, \quad 0 \leq h < T + s_0.$$

Thus

$$\|\int_0^{T+s_0} U(T+s_0, h) f(h, u(h, \phi), u_h(\phi)) dh\|_\eta \leq M_1 M_2 T^{1-\gamma}/(1-\gamma), \quad \phi \in H.$$

Therefore

$$\{\int_0^{T+s_0} U(T + s_0, h) f(h, u(h, \phi), u_h(\phi)) dh : \phi \in H\} \tag{5.21}$$

is bounded in $\mathbb{X}_\eta$. Then use the fact that the embedding $\mathbb{X}_\eta \to \mathbb{X}$ is compact again, we see that the set defined by (5.21) is precompact in $\mathbb{X}$. Now the Ascoli-Arzela theorem implies that the map P is a compact operator. □

Similar to the case for equations without delay, we have

Theorem 5.11. *Let the assumptions 5.1 and 5.2 be satisfied and let $T > r$. If the solutions of Eq. (5.14) are ultimate bounded, then Eq. (5.14) has a T-periodic solution.*

Theorem 5.12. *Assume that there exist functions ("wedges") $W_i, i = 1, 2, 3$, with $W_i : [0, \infty) \to [0, \infty), W_i(0) = 0, W_i$ strictly increasing, and $W_1(t) \to \infty, t \to \infty$. Further, assume that there exists a (Liapunov) function $V : \mathbb{X} \to \Re$ (reals) such that for some constant $M > 0$, when u is a solution of Eq. (5.14) with $\|u(t)\| \geq M$, then*

(a). $W_1(\|u(t)\|) \leq V(u(t)) \leq W_2(\|u(t)\|)$, *and*
(b). $\frac{d}{dt}V(u(t)) \leq -W_3(\|u(t)\|)$.

Then solutions of Eq. (5.14) are bounded and ultimate bounded.

Exercise 21.

(1) Prove Theorem 5.7.
(2) Prove Lemma 5.4.
(3) Prove Theorem 5.8.
(4) In the proof of Theorem 5.10, show that the Poincaré map P defined in (5.16) is continuous.
(5) In the proof of Theorem 5.10, verify that as functions on $\cdot \in [-r, 0]$,

$$\{U(T+\cdot,0)\phi(0) : \phi \in H\} = \{U(T+\cdot,k)\Big[U(k,0)\phi(0)\Big] : \phi \in H\}$$

given in (5.18) is equicontinuous.
(6) Prove Theorem 5.11.
(7) Prove Theorem 5.12.

5.1.3 *Nonlinear Equations With Infinite Delay*

Consider the following nonlinear differential equation with infinite delay,

$$u'(t) + A(t)u(t) = f(t, u(t), u_t), \quad t > 0, \quad u(s) = \phi(s), \quad s \leq 0, \quad (5.22)$$

in a general Banach space $\mathbb{X}$, where $u_t(s) = u(t+s)$, $s \leq 0$.

The materials presented here are based on [Liu (62); Liu Naito and Minh (63)]. Similar to equations with finite delay, we now need to consider functions defined on $(-\infty, 0]$, where in general only a seminorm is available. In this regard, an approach using axioms for seminormed abstract spaces can be made, see Henriquez [Henriquez (45)] and Hale and Kato [Hale and Kato (41)]. Here, we will approach in a simpler way by considering functions in the continuous functions space $C((-\infty, 0], \mathbb{X})$ against a fixed function g. These spaces are called "weighted" (or "friendly" in some literature) phase space and are denoted by C_g.

To define such a phase space C_g for Eq. (5.22), we have

Lemma 5.5. *There exists an integer $K_0 > 1$ such that*

$$(\frac{1}{2})^{K_0 - 1} M_0 < 1, \quad (5.23)$$

where $M_0 = \sup_{t\in[0,T]} \|U(t,0)\|$ *is finite. Next, let* $w_0 = T/K_0$*, then there exists a function* g *on* $(-\infty, 0]$ *such that* $g(0) = 1$*,* $g(-\infty) = \infty$*,* g *is decreasing on* $(-\infty, 0]$*, and for* $d \geq w_0$ *one has*

$$\sup_{s\leq 0} \frac{g(s)}{g(s-d)} \leq \frac{1}{2}. \tag{5.24}$$

Proof. Such a function g exists, e.g., we can take $g(s) = e^{-as}$ where $a > 0$ is such that $e^{aw_0} \geq 2$. □

For the function g given in Lemma 5.5, define the continuous functions space

$$C_g = \Big\{\phi : \phi \in C((-\infty, 0], \mathbb{X}) \text{ and } \lim_{s\to-\infty} \frac{\|\phi(s)\|}{g(s)} = 0\Big\}. \tag{5.25}$$

Then C_g coupled with the norm

$$|\phi|_g = \sup_{s\leq 0} \frac{\|\phi(s)\|}{g(s)}, \quad \phi \in C_g, \tag{5.26}$$

is a Banach space, see an exercise.

Similar to the study of equations with finite delay, we have

Theorem 5.13. *Let the assumptions 5.1 and 5.2 be satisfied and let* $\phi \in C_g$*. Then there exists a constant* $\alpha > 0$ *such that Eq. (5.22) has a unique (mild) solution* $u : (-\infty, \alpha] \to \mathbb{X}$ *satisfying* $u_0 = \phi$ *(i.e.,* $u(s) = \phi(s)$*,* $s \leq 0$*), and*

$$u(t) = U(t,0)\phi(0) + \int_0^t U(t,h)f(h, u(h), u_h)dh, \quad t \in [0, \alpha]. \tag{5.27}$$

To derive periodic solutions, we assume that solutions exist on $[0, \infty)$. We will write $u = u(\cdot, \phi)$ for the unique solution with the initial function ϕ.

Now, for equations with infinite delay, the Poincaré operator will map an element ϕ of $C((-\infty, 0], \mathbb{X})$ to an element of $C((-\infty, 0], \mathbb{X})$, given by

$$P\phi = u_T(\cdot, \phi), \quad \phi \in C((-\infty, 0], \mathbb{X}), \tag{5.28}$$

or, for $s \in (-\infty, 0]$,

$$(P\phi)(s) = u_T(s, \phi) = u(T + s, \phi)$$

$$= \begin{cases} U(T+s, 0)\phi(0) + \displaystyle\int_0^{T+s} U(T+s, h)f(h, u(h), u_h)dh, \\ \hfill T + s > 0, \\ \phi(T+s), \quad T + s \leq 0. \end{cases}$$

Similar to Lemma 5.4, we have

Lemma 5.6. *Let the assumptions 5.1 and 5.2 be satisfied.*

(a). If $u(t)$ is a solution of Eq. (5.22), then so is $u(t+T)$, $t \geq 0$.
(b). Equation (5.22) has a T-periodic solution if and only if the Poincaré map $P : C((-\infty,0],\mathbb{X}) \to C((-\infty,0],\mathbb{X})$ defined in (5.28) has a fixed point.

Definitions concerning boundedness and ultimate boundedness given in the previous sections can be revised as follows to suit Eq. (5.22).

Definition 5.6. The solutions of Eq. (5.22) are said to be **bounded** if for each $B_1 > 0$ there is a $B_2 = B_2(B_1) > 0$ such that $\{|\phi|_g \leq B_1, t \geq 0\}$ implies $\|u(t,\phi)\| < B_2$.

Definition 5.7. The solutions of Eq. (5.22) are said to be **ultimate bounded** if there is a bound $B > 0$ such that for each $B_3 > 0$, there is a $K = K(B, B_3) > 0$ such that $\{|\phi|_g \leq B_3, t \geq K\}$ implies $\|u(t,\phi)\| < B$.

For the relationship between boundedness and ultimate boundedness, we also have the following result.

Theorem 5.14. *Assume that $f(t,u,v)$ in Eq. (5.22) is continuous and is Lipschitzian in u and in v. If the solutions of Eq. (5.22) are ultimately bounded, then they are also bounded.*

For equations without delay and with finite delay, we obtained periodic solutions because we were able to prove the compactness of the Poincaré operators for those equations, therefore compact sets could be constructed and Horn's fixed point theorem could be applied to derive fixed points and hence periodic solutions.

Now, for Eq. (5.22) with infinite delay, the geometry of the Poincaré operator defined in (5.28) is that the function ϕ defined from 0 all the way to the left $(-\infty)$ is now mapped to a function from $T(>0)$ all the way to the left. Thus, under the Poincaré operator, the history of the initial function ϕ on $(-\infty,0]$ is carried over to become the part of $P\phi$ from 0 all the way to the left. Now, to obtain the compactness of the Poincaré operator, we need to use the Ascoli-Arzela theorem and verify that the solution set (for the initial functions in a bounded set) is precompact for every $s \leq T$. This is impossible now because when $s \leq 0$, the solution set is the same as the initial functions, which may be arbitrary, or "bad", i.e., noncompact.

That is, it is possible that under the Poincaré operator P defined in (5.28), a bounded set gets mapped into a noncompact set. Therefore, the Poincaré operator P defined in (5.28) is not compact. Hence, all the fixed point theorems requiring compactness, such as Browder's, Horn's, Schauder's, and Schauder-Tychonov's, are not applicable now for equations with infinite delay in general Banach spaces.

This causes the major difficulty for the study of periodic solutions of infinite delay equations in general Banach spaces. To overcome this difficulty, one way is to use Kuratowski's measure of *non-compactness* for condensing operators to get fixed points and hence periodic solutions, where by the name of "non-compactness", the compactness requirement is removed. Let's introduce these notions.

Definition 5.8. The **Kuratowski's measure of non-compactness** (or the α measure) for a bounded set H of a Banach space Y is defined as

$$\alpha(H) = \inf \Big\{ d > 0 : H \text{ has a finite cover of diameter } < d \Big\}. \quad (5.29)$$

We need to use the following basic properties of the α measure here.

Lemma 5.7. *[Lakshmikantham and Leela (52)] Let A and B be bounded sets of a Banach space Y. Then*

(1) $\alpha(A) \leq dia(A)$. $(dia(A) = \sup\{|x-y|_Y : x, y \in A\}.)$
(2) $\alpha(A) = 0$ if and only if A is precompact.
(3) $\alpha(\lambda A) = |\lambda|\alpha(A),\ \lambda \in \Re$. $(\lambda A = \{\lambda x : x \in A\})$
(4) $\alpha(A \cup B) = \max\{\alpha(A), \alpha(B)\}$.
(5) $\alpha(A + B) \leq \alpha(A) + \alpha(B)$. $(A + B = \{x + y : x \in A, y \in B\})$
(6) $\alpha(A) \leq \alpha(B)$ if $A \subseteq B$.

Lemma 5.8. *Let A with norm $|\cdot|_A$ and C with norm $|\cdot|_C$ be bounded. If there is a surjective map $Q : C \to A$ such that for any $c, d \in C$ one has $|Q(c) - Q(d)|_A \leq |c - d|_C$, then $\alpha(A) \leq \alpha(C)$.*

Proof. For any $\varepsilon > 0$, there exist bounded sets $G^i \subseteq C$, $i = 1, ..., m$, such that

$$dia(G^i) \leq \alpha(C) + \varepsilon, \quad C = \cup_{i=1}^m G^i. \quad (5.30)$$

Now, Q is surjective, so that $A = \cup_{i=1}^m Q(G^i)$. And for $a, b \in Q(G^i)$ we may assume that $a = Q(c)$, $b = Q(d)$ for some $c, d \in G^i$. Thus

$$\begin{aligned} |a - b|_A = |Q(c) - Q(d)|_A \leq |c - d|_C \leq dia(G^i) \\ \leq \alpha(C) + \varepsilon. \end{aligned} \quad (5.31)$$

This implies $dia(Q(G^i)) \leq \alpha(C) + \varepsilon$, and hence from Lemma 5.7 (i), $\alpha(Q(G^i)) \leq dia(Q(G^i)) \leq \alpha(C) + \varepsilon$. Therefore Lemma 5.7 (iv) implies that $\alpha(A) \leq \alpha(C) + \varepsilon$. Since $\varepsilon > 0$ is arbitrary, the result is true. □

Definition 5.9. An operator P is said to be a **condensing operator** on a Banach space Y if P is continuous and takes a bounded set into a bounded set, and $\alpha(P(B)) < \alpha(B)$ for every bounded set B of Y with $\alpha(B) > 0$, (where $\alpha(\cdot)$ is the α measure).

For condensing operators, the following asymptotic fixed point theorem from Hale and Lunel [Hale and Lunel (42)] is very useful. It is similar to Horn's and Browder's asymptotic fixed point theorems but doesn't require compactness.

Theorem 5.15. *(Hale and Lunel [Hale and Lunel (42)]) Suppose $S_0 \subseteq S_1 \subseteq S_2$ are convex bounded subsets of a Banach space Y, S_0 and S_2 are closed, and S_1 is open in S_2, and suppose $P : S_2 \to Y$ is (S_2)-condensing in the following sense: if U and $P(U)$ are contained in S_2 and $\alpha(U) > 0$, then $\alpha(P(U)) < \alpha(U)$. If $P^j(S_1) \subseteq S_2$, $j \geq 0$, and, for any compact set $H \subseteq S_1$, there is a number $N(H)$ such that $P^k(H) \subseteq S_0$, $k \geq N(H)$, then P has a fixed point.*

Based on Theorem 5.15, we have

Theorem 5.16. *Suppose $S_0 \subseteq S_1 \subseteq S_2$ are convex bounded subsets of a Banach space Y, S_0 and S_2 are closed, and S_1 is open in S_2, and suppose P is a condensing operator in Y. If $P^j(S_1) \subseteq S_2$, $j \geq 0$, and there is a number $N(S_1)$ such that $P^k(S_1) \subseteq S_0$, $k \geq N(S_1)$, then P has a fixed point.*

The advantage of these is that the compactness requirement is removed, as long as the operator shrinks sets under α measure. We will apply the fixed point theorem 5.16 to Eq. (5.22) with infinite delay in C_g and derive periodic solutions using ultimate boundedness.

For the Banach space C_g defined above, we have

Lemma 5.9. *Let u be a continuous function on $(-\infty, T]$ such that $|u_t|_g$ is finite for every $t \in [0, T]$. Then for any $0 \leq h < r \leq T$ with $r - h \geq w_0$ (w_0 is from Lemma 5.5), one has*

$$|u_r|_g \leq \max\Big\{ \sup_{s\in[h,r]} \|u(s)\|, \ \frac{1}{2}|u_h|_g \Big\}. \tag{5.32}$$

Proof. We have

$$
\begin{aligned}
|u_r|_g &= \sup_{s\le 0}\frac{\|u_r(s)\|}{g(s)} = \sup_{s\le 0}\frac{\|u(r+s)\|}{g(s)}\\
&= \sup_{l\le r}\frac{\|u(l)\|}{g(l-r)} \qquad (r+s=l)\\
&\le \max\left\{\sup_{l\in[h,r]}\|u(l)\|\frac{1}{g(l-r)},\ \sup_{l\le h}\frac{\|u(l)\|}{g(l-r)}\right\}\\
&= \max\left\{\sup_{l\in[h,r]}\|u(l)\|\frac{1}{g(l-r)},\ \sup_{l\le h}\frac{\|u(l)\|}{g(l-h)}\frac{g(l-h)}{g(l-r)}\right\}\\
&\le \max\left\{\sup_{l\in[h,r]}\|u(l)\|,\ \sup_{s\le 0}\frac{\|u(h+s)\|}{g(s)}\frac{g(s)}{g(s-(r-h))}\right\} \quad (l-h=s)\\
&\le \max\left\{\sup_{s\in[h,r]}\|u(s)\|,\ \frac{1}{2}|u_h|_g\right\} \qquad (5.33)
\end{aligned}
$$

by using Lemma 5.5. □

To estimate the solutions, we have

Lemma 5.10. *Let the assumptions 5.1 and 5.2 be satisfied and let u and y be two solutions of Eq. (5.22) (with initial functions u_0 and y_0 respectively) on $(-\infty, L]$, $L>0$. Then for $t\in[0,L]$,*

$$|u_t - y_t|_g \le K_1|u_0 - y_0|_g e^{K_2 t}, \tag{5.34}$$

where K_1 and K_2 are some constants.

Proof. Similar to the proof of Lemma 5.9, we have, for $t\in[0,L]$,

$$|u_t - y_t|_g \le \max\left\{\sup_{s\in[0,t]}\|u(s)-y(s)\|,\ |u_0-y_0|_g\right\}. \tag{5.35}$$

Next, using Lipschitz conditions, we may assume that

$$\|f(h,u(h),y_h) - f(h,y(h),y_h)\| \le k_0\|u(h)-y(h)\|, \tag{5.36}$$

$$\|f(h,u(h),u_h) - f(h,u(h),y_h)\| \le k_1|u_h - y_h|_g, \tag{5.37}$$

for some constants k_0 and k_1. Let $M_0 = \sup_{t\in[0,T]} \|U(t,0)\|$, $M_1 = \sup_{0\le h\le s\le T} \|U(s,h)\|$, which are finite. Then for $s \in [0,t]$,

$$\begin{aligned}
\|u(s) - y(s)\| &= \|U(s,0)(u(0) - y(0)) \\
&\quad + \int_0^s U(s,h)\Big[f(h,u(h),u_h) - f(h,y(h),y_h)\Big]dh\| \\
&= \|U(s,0)(u(0) - y(0)) \\
&\quad + \int_0^s U(s,h)\Big[f(h,u(h),u_h) - f(h,u(h),y_h)\Big]dh\| \\
&\quad + \int_0^s U(s,h)\Big[f(h,u(h),y_h) - f(h,y(h),y_h)\Big]dh\| \\
&\le M_0\|u(0) - y(0)\| + \int_0^s M_1 k_0 \|u(h) - y(h)\| dh \\
&\quad + \int_0^s M_1 k_1 |u_h - y_h|_g dh \\
&\le M_0 |u_0 - y_0|_g + \int_0^t M_1(k_0 + k_1)|u_h - y_h|_g dh. \qquad (5.38)
\end{aligned}$$

Thus, from (5.35), we have

$$\begin{aligned}
|u_t - y_t|_g &\le \sup_{s\in[0,t]} \|u(s) - y(s)\| + |u_0 - y_0|_g \\
&\le (M_0 + 1)|u_0 - y_0|_g + \int_0^t M_1(k_0 + k_1)|u_h - y_h|_g dh.
\end{aligned}$$

Now, the Gronwall's inequality implies (5.34). □

An immediate consequence of Lemma 5.10 is the following local boundedness property of the solutions.

Theorem 5.17. *Let the assumptions 5.1 and 5.2 be satisfied and let $D \subset C_g$ be bounded. Then for any $L > 0$, solutions of Eq. (5.22) with initial functions in D are bounded on $[0, L]$. That is, there exists a constant $E = E(D, L) > 0$ such that if $u(\cdot) = u(\cdot, \phi)$ with $\phi \in D$, then $\|u(t)\| \le E$ for $t \in [0, L]$.*

Proof. Let $y = y(\phi_0)$ be a fixed solution with $\phi_0 \in D$. Then $|u_L|_g \leq |u_L - y_L|_g + |y_L|_g$, and hence Lemma 5.10 implies that $\{|u_L(\phi)|_g : \phi \in D\}$ is bounded. Therefore the result is true by using the definition of the norm in C_g. □

Next, for $D \subset C_g$ and $u(\phi)$ the unique solution with $u_0(\phi) = \phi$, we define $W_l(D) = \{u_l(\phi) : \phi \in D\}$ and $W_{[h,r]}(D) = \{u_{[h,r]}(\phi) : \phi \in D\}$, where $u_{[h,r]}$ means the restriction of u on $[h,r]$.

Lemma 5.11. *Let the assumptions 5.1 and 5.2 be satisfied. If $D \subset C_g$ is bounded, then $W_{[0,T]}(D) \subset C([0,T],\mathbb{X})$ is bounded and $W_r(D) \subset C_g$ is bounded for each $r \in [0,T]$. And for any $0 \leq h < r \leq T$ with $r - h \geq w_0$ (w_0 is from Lemma 5.5), one has*

$$\alpha(W_r(D)) \leq \max\left\{\alpha(W_{[h,r]}(D)),\ \frac{1}{2}\alpha(W_h(D))\right\}. \tag{5.39}$$

Proof. First, Theorem 5.17 implies that $W_{[0,T]}(D) \subset C([0,T],\mathbb{X})$ is bounded. This result and Lemma 5.9 (with $h = 0$) imply that for each $r \in [0,T]$, $W_r(D)$ is bounded in C_g. Now, for any $\varepsilon > 0$, there exist bounded sets $P^i \subseteq W_{[h,r]}(D)$, $i = 1, ..., m$, and bounded sets $Q^j \subseteq W_h(D)$, $j = 1, ..., n$, such that

$$dia(P^i) \leq \alpha(W_{[h,r]}(D)) + \varepsilon,\ \ W_{[h,r]}(D) = \cup_{i=1}^m P^i, \tag{5.40}$$

$$dia(Q^j) \leq \alpha(W_h(D)) + 2\varepsilon, \qquad W_h(D) = \cup_{j=1}^n Q^j. \tag{5.41}$$

Put

$$Y_r^{i,j} = \left\{u_r \in W_r(D) : u_{[h,r]} \in P^i,\ u_h \in Q^j\right\}. \tag{5.42}$$

Then we have

$$W_r(D) = \cup_{i=1}^m \cup_{j=1}^n Y_r^{i,j}. \tag{5.43}$$

For each $Y_r^{i,j}$, if $u_r, w_r \in Y_r^{i,j}$, then from the proof of Lemma 5.9,

$$\begin{aligned}
|u_r - w_r|_g &\leq \max\left\{\sup_{s\in[h,r]} \|u(s) - w(s)\|,\ \frac{1}{2}|u_h - w_h|_g\right\} \\
&\leq \max\left\{dia(P^i),\ \frac{1}{2}dia(Q^j)\right\} \\
&\leq \max\left\{\alpha(W_{[h,r]}(D)) + \varepsilon,\ \frac{1}{2}(\alpha(W_h(D)) + 2\varepsilon)\right\} \\
&= \max\left\{\alpha(W_{[h,r]}(D)),\ \frac{1}{2}\alpha(W_h(D))\right\} + \varepsilon.
\end{aligned} \tag{5.44}$$

This implies, using Lemma 5.7 (i), that

$$\alpha(Y_r^{i,j}) \le dia(Y_r^{i,j}) \le \max\Big\{\alpha(W_{[h,r]}(D)),\ \frac{1}{2}\alpha(W_h(D))\Big\} + \varepsilon. \quad (5.45)$$

Then Lemma 5.7 (iv) implies that

$$\alpha(W_r(D)) \le \max\Big\{\alpha(W_{[h,r]}(D)),\ \frac{1}{2}\alpha(W_h(D))\Big\} + \varepsilon. \quad (5.46)$$

Since $\varepsilon > 0$ is arbitrary, the result is true. □

By using the Ascoli-Arzela theorem, we have the following result. The idea of the proof is similar to the case for equations with finite delay.

Lemma 5.12. *Let the assumptions 5.1 and 5.2 be satisfied and let $D \subset C_g$ be bounded. Then $\alpha(W_{[l,r]}(D)) = 0$ for any $0 < l < r \le T$.*

Proof. By Lemma 5.7 (ii), we need to prove that the Ascoli-Arzela theorem can be applied to the bounded set $E = W_{[l,r]}(D) \subset C([l,r],\mathbb{X})$.

Note that a function in E can be expressed as, for $s \in [l,r]$,

$$u(s,\phi) = U(s,0)\phi(0) + \int_0^s U(s,h)f(h,u(h),u_h)dh,\ \phi \in D. \quad (5.47)$$

Since $l > 0$, there is $k > 0$ such that $s > k$ for $s \in [l,r]$. For $s \in [l,r]$, one has

$$U(s,0)\phi(0) = U(s,k)U(k,0)\phi(0), \quad \phi \in D. \quad (5.48)$$

Fix $\eta \in (0,1)$. Then from Lemma 5.3 (i), $U(k,0) : \mathbb{X} \to \mathbb{X}_\eta$ is bounded. Next the embedding $\mathbb{X}_\eta \to \mathbb{X}$ is compact, thus $\{U(k,0)\phi(0) : \phi \in D\}$ is precompact in $\mathbb{X}$ since $\{\phi(0) : \phi \in D\}$ is bounded in $\mathbb{X}$. Therefore, the closure of $\{U(k,0)\phi(0) : \phi \in D\}$ is compact in $\mathbb{X}$. Now, one can verify that as functions on $\cdot \in [l,r]$,

$$\{U(\cdot,0)\phi(0) : \phi \in D\} = \{U(\cdot,k)\Big[U(k,0)\phi(0)\Big] : \phi \in D\} \quad (5.49)$$

is equicontinuous. Next, from Lemma 5.3 (ii), for $0 \le \gamma < 1$, there is a constant $C(\gamma)$, such that for $s_1,\ s_2 \in [l,r]$,

$$\|\int_0^{s_2} U(s_2,h)f(h,u(h),u_h)dh - \int_0^{s_1} U(s_1,h)f(h,u(h),u_h)dh\|$$

$$\le C(\gamma)|s_1 - s_2|^\gamma \max_{0\le h\le T} \|f(h,u(h),u_h)\|. \quad (5.50)$$

By using Lemma 5.11, we see that the variables in f are bounded. Now f maps a bounded set into a bounded set, thus there exists $M_2 = M_2(D) > 0$ such that

$$\|f(t,u(t),u_t(\phi))\| \le M_2, \quad t \in [0,T], \quad \phi \in D.$$

Therefore, as functions on $\cdot \in [l,r]$,

$$\{\int_0^{\cdot} U(\cdot,h)f(h,u(h),u_h)dh : \phi \in D\} \tag{5.51}$$

is also equicontinuous. Therefore, functions in E are equicontinuous.

In the following, to check the precompactness of the functions at every point of $[l,r]$, we fix $s_0 \in [l,r]$. From the above arguments, we also know that

$$\{U(s_0,0)\phi(0) : \phi \in D\} \tag{5.52}$$

is precompact in $\mathbb{X}$. Next, for $\phi \in D$, we let $g(t) = f(t,u(t),u_t(\phi))$. Then

$$\int_0^{s_0} U(s_0,h)f(h,u(h),u_h(\phi))dh \equiv K(0,g)(s_0) \in \mathbb{X}_\eta$$

according to Lemma 5.3 (iii). Also note that by Lemma 5.3 (i), there are constants $\gamma \in (0,1)$ and $M_3 > 0$ such that

$$\|U(s_0,h)\|_{0,\eta} \le M_3(s_0-h)^{-\gamma}, \quad 0 \le h < s_0.$$

Thus

$$\|\int_0^{s_0} U(s_0,h)f(h,u(h),u_h(\phi))dh\|_\eta \le M_3M_2T^{1-\gamma}/(1-\gamma), \quad \phi \in D.$$

Therefore

$$\{\int_0^{s_0} U(s_0,h)f(h,u(h),u_h(\phi))dh : \phi \in D\} \tag{5.53}$$

is bounded in $\mathbb{X}_\eta$. Then use the fact that the embedding $\mathbb{X}_\eta \to \mathbb{X}$ is compact again, we see that the set defined by (5.53) is precompact in $\mathbb{X}$. Now the Ascoli-Arzela theorem implies that E is precompact. Thus by Lemma 5.7 (ii), $\alpha(W_{[l,r]}(D)) = \alpha(E) = 0$. □

Next, we prove that the operator P defined in (5.28) is condensing in C_g.

Theorem 5.18. *Let the assumptions 5.1 and 5.2 be satisfied. Then the operator P defined in (5.28) is condensing in C_g with g given in Lemma 5.5.*

Proof. From Lemma 5.10, we have

$$|P(\phi) - P(\varphi)|_g = |u_T(\phi) - u_T(\varphi)|_g \le K_1 e^{K_2 T}|\phi - \varphi|_g, \tag{5.54}$$

thus P is continuous, and takes a bounded set into a bounded set. Next, let $D \subset C_g$ be bounded with $\alpha(D) > 0$. By using Lemmas 5.11 and 5.12 repeatedly, we have (w_0 is from Lemma 5.5)

$$\begin{aligned}
\alpha(P(D)) &= \alpha(W_T(D)) \leq \max\Big\{\alpha(W_{[T-w_0,T]}(D)),\ \frac{1}{2}\alpha(W_{T-w_0}(D))\Big\} \\
&= \frac{1}{2}\alpha(W_{T-w_0}(D)) \\
&\leq \frac{1}{2}\max\Big\{\alpha(W_{[T-2w_0,T-w_0]}(D)),\ \frac{1}{2}\alpha(W_{T-2w_0}(D))\Big\} \\
&= (\frac{1}{2})^2\alpha(W_{T-2w_0}(D)) \\
&\leq (\frac{1}{2})^2\max\Big\{\alpha(W_{[T-3w_0,T-2w_0]}(D)),\ \frac{1}{2}\alpha(W_{T-3w_0}(D))\Big\} \\
&= (\frac{1}{2})^3\alpha(W_{T-3w_0}(D)) \\
&\cdots\cdots \\
&\leq (\frac{1}{2})^{K_0-1}\max\Big\{\alpha(W_{[0,T-(K_0-1)w_0]}(D)),\ \frac{1}{2}\alpha(D)\Big\}. \qquad (5.55)
\end{aligned}$$

Next, for $\cdot \in [0, T-(K_0-1)w_0]$,

$$\begin{aligned}
W_{[0,T-(K_0-1)w_0]}(D) \subseteq &\Big\{U(\cdot,0)\phi(0) : \phi \in D\Big\} \\
&+\Big\{\int_0^{\cdot} U(\cdot,h)f(h,u(h),u_h(\phi))dh : \phi \in D\Big\}.
\end{aligned}$$

And for $t \in [0, T-(K_0-1)w_0]$,

$$\begin{aligned}
\|U(t,0)\phi(0) - U(t,0)\varphi(0)\| &= \|U(t,0)(\phi(0)-\varphi(0))\| \\
&\leq M_0\|\phi(0)-\varphi(0)\| \leq M_0|\phi-\varphi|_g,
\end{aligned}$$

where $M_0 = \sup_{t\in[0,T]}\|U(t,0)\|$, then we have from Lemma 5.7 (iii) and Lemma 5.8 that (for $\cdot \in [0, T-(K_0-1)w_0]$)

$$\alpha\{U(\cdot,0)\phi(0) : \phi \in D\} \leq M_0\alpha(D). \qquad (5.56)$$

Similar to the proof in Lemma 5.12 we see that for $\cdot \in [0, T-(K_0-1)w_0]$,

$$\alpha\{\int_0^{\cdot} U(\cdot,h)f(h,u(h),u_h(\phi))dh : \phi \in D\} = 0. \qquad (5.57)$$

Therefore we have from Lemma 5.7 (v) that

$$\alpha(W_{[0,T-(K-1)w_0]}(D)) \leq M_0\alpha(D). \qquad (5.58)$$

Thus, from Lemma 5.5, (5.55), and (5.58), we have (note that $M_0 = \sup_{t\in[0,T]} \|U(t,0)\| \geq 1$)

$$\alpha(P(D)) \leq (\frac{1}{2})^{K_0-1} \max\left\{M_0\alpha(D), \ \frac{1}{2}\alpha(D)\right\}$$
$$\leq (\frac{1}{2})^{K_0-1} M_0\alpha(D) < \alpha(D). \tag{5.59}$$

This proves that the operator P is condensing in C_g. □

Now, we are ready to prove the existence of periodic solutions for infinite delay differential equations in general Banach spaces.

Theorem 5.19. *Let assumptions 5.1 and 5.2 be satisfied. If the solutions of Eq. (5.22) are ultimate bounded, then Eq. (5.22) has a T-periodic solution.*

Proof. From Theorem 5.14, the solutions of Eq. (5.22) are also bounded. Let the operator P be defined in (5.28). Similar to the study of equations without delay and with finite delay, we have

$$P^m(\phi) = u_{mT}(\phi), \ \ \phi \in C_g, \ \ m = 1, 2, \cdots. \tag{5.60}$$

Next, let $B > 0$ be the bound in the definition of ultimate boundedness. Using boundedness, there is a $B_1 > B$ such that $\{|\phi|_g \leq B, t \geq 0\}$ implies $\|u(t,\phi)\| < B_1$. Also, there is a $B_2 > B_1$ such that $\{|\phi|_g \leq B_1, t \geq 0\}$ implies $\|u(t,\phi)\| < B_2$. Next, using ultimate boundedness, there is a positive integer J such that $\{|\phi|_g \leq B_1, t \geq JT\}$ implies $\|u(t,\phi)\| < B$.

Now let

$$S_2 \equiv \{\phi \in C_g : |\phi|_g \leq B_2\},$$
$$W \equiv \{\phi \in C_g : |\phi|_g < B_1\}, \ S_1 \equiv W \cap S_2, \tag{5.61}$$
$$S_0 \equiv \{\phi \in C_g : |\phi|_g \leq B\},$$

so that $S_0 \subseteq S_1 \subseteq S_2$ are convex bounded subsets of Banach space C_g, S_0

and S_2 are closed, and S_1 is open in S_2. Next, for $\phi \in S_1$ and $j \geq 0$,

$$
\begin{aligned}
|P^j \phi|_g = |u_{jT}(\phi)|_g &= \sup_{s \leq 0} \frac{\|u_{jT}(s)\|}{g(s)} = \sup_{s \leq 0} \frac{\|u(jT+s)\|}{g(s)} \\
&\leq \max\Big\{ \sup_{s \leq -jT} \frac{\|u(jT+s)\|}{g(s)}, \sup_{s \in [-jT,0]} \frac{\|u(jT+s)\|}{g(s)} \Big\} \\
&\leq \max\Big\{ \sup_{l \leq 0} \frac{\|u(l)\|}{g(l-jT)}, \sup_{l \in [0,jT]} \|u(l)\| \Big\} \\
&\leq \max\Big\{ \sup_{l \leq 0} \frac{\|u(l)\|}{g(l)}, \sup_{l \in [0,jT]} \|u(l)\| \Big\} \\
&\leq \max\Big\{ |\phi|_g, \ B_2 \Big\} \leq B_2,
\end{aligned} \tag{5.62}
$$

which implies $P^j(S_1) \subseteq S_2$, $j \geq 0$. Now, we prove that there is a number $N(S_1)$ such that $P^k(S_1) \subseteq S_0$ for $k \geq N(S_1)$. To this end, we choose a positive integer $m = m(B_1)$ such that

$$
(\frac{1}{2})^m < \frac{B}{B_1}, \tag{5.63}
$$

and then choose an integer $N = N(S_1) > J$ such that

$$
NT > m w_0 \quad \text{and} \quad \frac{B_2}{g(-(N-J)T)} < B, \tag{5.64}
$$

where w_0 is from Lemma 5.5. Then for $\phi \in S_1$ and $k \geq N$,

$$
\begin{aligned}
|P^k \phi|_g = |u_{kT}(\phi)|_g &= \sup_{s \leq 0} \frac{\|u_{kT}(s)\|}{g(s)} = \sup_{s \leq 0} \frac{\|u(kT+s)\|}{g(s)} \\
\leq \max\Big\{ \sup_{s \leq -kT} \frac{\|u(kT+s)\|}{g(s)}, &\sup_{s \in [-kT,-(k-J)T]} \frac{\|u(kT+s)\|}{g(s)}, \\
&\sup_{s \in [-(k-J)T,0]} \frac{\|u(kT+s)\|}{g(s)} \Big\}.
\end{aligned} \tag{5.65}
$$

For the terms in (5.65), we have

$$
\sup_{s \in [-(k-J)T,0]} \frac{\|u(kT+s)\|}{g(s)} \leq \sup_{l \in [JT,kT]} \|u(l)\| < B, \tag{5.66}
$$

and

$$\sup_{s\in[-kT,-(k-J)T]} \frac{\|u(kT+s)\|}{g(s)} \leq \sup_{l\in[0,JT]} \frac{\|u(l)\|}{g(l-kT)}$$

$$\leq \frac{B_2}{g(-(k-J)T)} \leq \frac{B_2}{g(-(N-J)T)} < B, \tag{5.67}$$

and

$$\sup_{s\leq -kT} \frac{\|u(kT+s)\|}{g(s)} = \sup_{l\leq 0} \frac{\|u(l)\|}{g(l-kT)}$$

$$= \sup_{l\leq 0} \frac{\|u(l)\|}{g(l)} \frac{g(l)}{g(l-kT)}$$

$$\leq |\phi|_g \sup_{l\leq 0} \frac{g(l)}{g(l-kT)}$$

$$\leq B_1 \sup_{l\leq 0} \frac{g(l)}{g(l-w_0)} \frac{g(l-w_0)}{g(l-2w_0)} \cdots$$

$$\cdots \frac{g(l-(m-1)w_0)}{g(l-mw_0)} \frac{g(l-mw_0)}{g(l-kT)}. \tag{5.68}$$

Now, from Lemma 5.5, for $i \geq 0$,

$$\sup_{l\leq 0} \frac{g(l-iw_0)}{g(l-(i+1)w_0)} = \sup_{s\leq -iw_0} \frac{g(s)}{g(s-w_0)}$$

$$\leq \sup_{s\leq 0} \frac{g(s)}{g(s-w_0)} \leq \frac{1}{2}. \tag{5.69}$$

Thus, (5.68) becomes

$$\sup_{s\leq -kT} \frac{\|u(kT+s)\|}{g(s)} \leq B_1 (\frac{1}{2})^m \sup_{l\leq 0} \frac{g(l-mw_0)}{g(l-kT)}$$

$$< B_1 \frac{B}{B_1} \sup_{l\leq 0} \frac{g(l-mw_0)}{g(l-NT)} \leq B \sup_{l\leq 0} \frac{g(l-mw_0)}{g(l-mw_0)} = B. \tag{5.70}$$

Therefore, (5.65) becomes

$$|P^k\phi|_g \leq B, \ \ k \geq N, \tag{5.71}$$

which implies $P^k(S_1) \subseteq S_0$, $k \geq N(S_1)$. Now, Theorem 5.16 can be used to obtain a fixed point for the operator P, which, from Lemma 5.6, gives rise to a T-periodic solution of Eq. (5.22). This proves the theorem. □

Theorem 5.20. *Assume that there exist functions ("wedges") $W_i, i = 1, 2, 3$, with $W_i : [0, \infty) \to [0, \infty), W_i(0) = 0, W_i$ strictly increasing, and $W_1(t) \to \infty, t \to \infty$. Further, assume that there exists a (Liapunov) function $V : \mathbb{X} \to \Re$ (reals) such that for some constant $M > 0$, when u is a solution of Eq. (5.22) with $\|u(t)\| \geq M$, then*

(a). $W_1(\|u(t)\|) \leq V(u(t)) \leq W_2(\|u(t)\|)$, *and*
(b). $\frac{d}{dt}V(u(t)) \leq -W_3(\|u(t)\|)$.

Then solutions of Eq. (5.22) are bounded and ultimate bounded.

Exercise 22.

(1) Verify that C_g coupled with the norm defined in (5.26) is a Banach space.
(2) Prove Theorem 5.13.
(3) Prove Lemma 5.6.
(4) Prove Theorem 5.14.
(5) Prove Theorem 5.16.
(6) Verify (5.43).
(7) In the proof of Lemma 5.12, verify that as functions on $\cdot \in [l, r]$,

$$\{U(\cdot, 0)\phi(0) : \phi \in D\} = \{U(\cdot, k)\Big[U(k, 0)\phi(0)\Big] : \phi \in D\}$$

given in (5.49) is equicontinuous.
(8) In the proof of Theorem 5.19, verify (5.60).
(9) Prove Theorem 5.20.

5.1.4 *Non-Densely Defined Equations*

In the above studies of the existence of periodic solutions, the operator $A(t)$ is assumed to be densely defined, which is the case if we look at partial differential equations in L^p spaces. However, if we use the sup-norm to measure continuous functions, the corresponding operators may be non-densely defined.

Example 5.1. Consider the partial differential equation

$$\begin{cases} \frac{\partial}{\partial t}u(t, x) = \frac{\partial^2}{\partial x^2}u(t, x) + f(t, x), \quad (t, x) \in (0, \infty) \times (0, 1), \\ u(t, 0) = u(t, 1) = 0, \ t \geq 0, \\ u(0, x) = \Phi(x), \ x \in [0, 1]. \end{cases} \tag{5.72}$$

If we study Eq. (5.72) in $C[0,1]$ (the space of all continuous functions on $[0,1]$ with the sup-norm), and define

$$Au = u'', \quad D(A) = \{u \in C^2[0,1] : u(0) = u(1) = 0\}. \tag{5.73}$$

Then the closure of $D(A)$ is

$$\overline{D(A)} = \{u \in C[0,1] : u(0) = u(1) = 0\} \quad \neq \quad C[0,1], \tag{5.74}$$

thus A is not densely defined on $C[0,1]$.

In the following, we will look at

$$u'(t) = Au(t) + f(t, u(t), u_t), \quad t > 0, \quad u_0 = \phi \in C\left([-r,0], \mathbb{X}\right), \tag{5.75}$$

where the linear operator A is non-densely defined and satisfies the Hille-Yosida condition. And we will present some results of [Ezzinbi and Liu (31)] without details.

Since now the operator A is non-densely defined, the semigroup theory cannot be used. Therefore, we will apply the *integrated semigroup theory.* For differential equations with finite delay, the wellposedness is established in the space

$$C_0 = \left\{\phi \in C\left([-r,0], \mathbb{X}\right) : \phi(0) \in \overline{D(A)}\right\},$$

so that C_0 will become the base space for this setting, replacing $C([-r,0], \mathbb{X})$.

The *part* A_0 of A in $\overline{D(A)}$ is defined by

$$A_0 = A \quad \text{on} \quad D(A_0) = \left\{x \in D(A) : Ax \in \overline{D(A)}\right\}.$$

Then it is known that the part A_0 of A generates a strongly continuous semigroup $S_0(\cdot)$ on $\overline{D(A)}$. Following [Liu (61)], we assume $T > r$ and we need to verify the compactness of the Poincaré operator $P : C_0 \to C_0$ defined by

$$P\phi = u_T(\cdot, \phi), \quad \phi \in C_0. \tag{5.76}$$

Assumption 5.3. Let $T > 0$ be a constant. The function f is continuous in all its variables, T-periodic in the first variable t and uniformly Lipschitzian in other variables.

Assumption 5.4. The semigroup $(S_0(t))_{t\geq 0}$ is compact on $\overline{D(A)}$. That means for each $t > 0$, the operator $S_0(t)$ is compact on $\overline{D(A)}$.

The definitions on boundedness and ultimate boundedness are the same as those for finite delay equations studied before. With some conditions, the approach of [Liu (61)] can be modified to construct some sets so that Horn's fixed point theorem can be applied to derive fixed points for the Poincaré operator defined in (5.76), and hence periodic solutions. We state these results as follows.

Theorem 5.21. *Assume that $f(t,u,v)$ in Eq. (5.75) is continuous and is Lipschitzian in u and in v. If the solutions of Eq. (5.75) are ultimately bounded, then they are also bounded.*

Theorem 5.22. *Let the assumptions 5.3 and 5.4 be satisfied and let $T > r$. If the solutions of Eq. (5.75) are bounded, then the Poincaré operator $P\phi = u_T(\cdot, \phi)$ on C_0 defined in (5.76) is compact.*

Theorem 5.23. *Let the assumptions 5.3 and 5.4 be satisfied and let $T > r$. If the solutions of Eq. (5.75) are ultimately bounded, then Eq. (5.75) has a T-periodic solution.*

Theorem 5.24. *Assume that there exist functions ("wedges") $W_i, i = 1,2,3$, with $W_i : [0,\infty) \to [0,\infty), W_i(0) = 0, W_i$ strictly increasing, and $W_1(t) \to \infty, t \to \infty$. Further, assume that there exists a (Liapunov) function $V : \mathbb{X} \to \Re$ (reals) such that for some constant $M > 0$, when u is a solution of Eq. (5.75) with $\|u(t)\| \geq M$, then*

(a). $W_1(\|u(t)\|) \leq V(u(t)) \leq W_2(\|u(t)\|)$, and
(b). $\frac{d}{dt}V(u(t)) \leq -W_3(\|u(t)\|)$.

Then solutions of Eq. (5.75) are bounded and ultimate bounded.

Exercise 23.

(1) In Example 5.1, verify that
$$\overline{D(A)} = \{u \in C[0,1] : u(0) = u(1) = 0\} \neq C[0,1].$$
(2) Study the wellposedness of Eq. (5.75) in
$$C_0 = \left\{\phi \in C\left([-r,0], \mathbb{X}\right) : \phi(0) \in \overline{D(A)}\right\}.$$
(3) Prove that the part A_0 of A generates a strongly continuous semigroup $T_0(\cdot)$ on $\overline{D(A)}$.
(4) Prove Theorem 5.21.
(5) Prove Theorem 5.22.
(6) Prove Theorem 5.23.
(7) Prove Theorem 5.24.

5.2 Evolution Semigroups and Almost Periodic Solutions

The studies of the previous sections can be extended to almost periodic solutions of nonlinear equations. The first question we are faced with is how to associate to a given nonlinear evolution equation evolution semigroups in suitable function spaces. The next one is to find common fixed points of these semigroups in chosen function spaces. It turns out that using evolution semigroups we can not only give simple proofs of some results in the finite dimensional case, but also extend them easily to the infinite dimensional case.

5.2.1 *Evolution Semigroups*

In this subsection we are mainly concerned with the existence of almost periodic solutions of evolution equations of the form

$$\frac{dx}{dt} = Ax + f(t, x, x_t) \tag{5.77}$$

where A is the infinitesimal generator of a C_0-semigroup $(T(t))_{t\geq 0}$ and f is a continuous operator from $\mathbb{R} \times \mathbb{X} \times C$ to $\mathbb{X}$. Note that our method used in this subsection applies to nonautonomous equations with almost periodic coefficients, not restricted to periodic or autonomous equations as in the previous sections.

Throughout this section we will denote by $C = BUC((-\infty, 0], \mathbb{X})$ the space of all uniformly continuous and bounded functions from $(-\infty, 0]$ to $\mathbb{X}$, and by x_t the map $x(t+\theta) = x_t(\theta)$, $\theta \in (-\infty, 0]$, where $x(\cdot)$ is defined on $(-\infty, a]$ for some $a > 0$.

In this subsection we will deal with evolution equations of the form

$$\frac{dx}{dt} = Ax + f(t, x) \ , \quad x \in \mathbb{X} \tag{5.78}$$

where $\mathbb{X}$ is a Banach space, A is the infinitesimal generator of a C_0-semigroup of linear operators $(S(t))_{t\geq 0}$ of type ω, i.e.

$$\|S(t)x - S(t)y\| \leq e^{\omega t}\|x - y\|, \ \forall\, t \geq 0, \ x, y \in \mathbb{X} \ ,$$

and f is a continuous operator from $\mathbb{R} \times \mathbb{X}$ to $\mathbb{X}$. Hereafter, recall that by a mild solution $x(t), t \in [s, \tau]$ of equation (5.78) we mean a continuous solution of the integral equation

$$x(t) = S(t-s)x + \int_s^t S(t-\xi)f(\xi, x(\xi)d\xi, \ \forall s \leq t \leq \tau. \tag{5.79}$$

Definition 5.10. (condition H4). Equation (5.78) is said to satisfy *condition H4* if

(1) A is the infinitesimal generator of a linear semigroup $(S(t))_{t\geq 0}$ of type ω in $\mathbb{X}$,
(2) f is a continuous operator from $\mathbb{R}\times\mathbb{X}$ to $\mathbb{X}$,
(3) There is a constant γ such that for every fixed $t\in\mathbb{R}$, the operator $(-f(t,\cdot)+\gamma I)$ is accretive in $\mathbb{X}$.

The following condition will be used frequently:

Definition 5.11. (condition H5). Equation (5.78) is said to satisfy *condition H5* if for every $u\in AP(\mathbb{X})$ the function $f(\cdot,u(\cdot))$ belongs to $AP(\mathbb{X})$ and the operator f_* taking u into $f(\cdot,u(\cdot))$ is continuous.

The main point of our study is to associate with equation (5.78) an evolution semigroup which plays a role similar to that of the monodromy operator for equations with periodic coefficients. Hereafter we will denote by $U(t,s)$, $t\geq s$, the evolution operator corresponding to equation (5.78) which satisfies the assumptions of Theorem A.30, i.e. $U(t,s)x$ is the unique solution of Eq. (5.79).

Proposition 5.1. *Let the conditions H4 and H5 be satisfied. Then with Eq. (5.78) one can associate an evolution semigroup $(T^h)_{h\geq 0}$ acting on $AP(\mathbb{X})$, defined as*

$$[T^hv](t)=U(t,t-h)v(t-h),\forall h\geq 0,\ t\in\mathbb{R},v\in AP(\mathbb{X}).$$

Moreover, this semigroup has the following properties:

(1) $T^h,h\geq 0$ is strongly continuous, and

$$T^hu=S^hu+\int_0^h S^{h-\xi}f_*(T^\xi u)d\xi,\ \forall h\geq 0,u\in AP(\mathbb{X}),$$

where $(S^hu)(t)=S(h)u(t-h),\forall h\geq 0,t\in\mathbb{R},u\in AP(\mathbb{X})$.

(2)

$$\|T^hu-T^hv\|\leq e^{(\omega+\gamma)h}\|u-v\|,\ \forall h\geq 0,u,v\in AP(\mathbb{X}).$$

Proof. We first look at the solutions to the equation

$$w(t)=S^{t-a}z+\int_a^t S^{t-\xi}f_*(w(\xi))d\xi\ \ \forall z\in AP(\mathbb{X}),t\geq a\in\mathbb{R}. \tag{5.80}$$

It may be noted that $(S^h)_{h\geq 0}$ is a strongly continuous semigroup of linear oparators in $AP(\mathbb{X})$ of type ω. Furthermore, for $\lambda>0,\lambda\gamma<1$ and $u,v\in$

$AP(\mathbb{X})$, from the accretiveness of the operators $-f(t,\cdot)+\gamma I$ we get

$$\begin{aligned}(1-\lambda\gamma)\|x-y\| &= (1-\lambda\gamma)\sup_t \|u(t)-v(t)\| \\ &= \sup_t (1-\lambda\gamma)\|u(t)-v(t)\| \\ &\le \sup_t \|u(t)-v(t)-\lambda[f(t,u(t))-f(t,v(t))]\| \\ &= \|u-v-\lambda(f_*u-f_*v)\|. \end{aligned} \tag{5.81}$$

This shows that $(-f_*+\gamma I)$ is accretive. In virtue of Theorem A.30 there exists a semigroup $(T^h)_{h\ge 0}$ such that

$$T^h u = S^h u + \int_0^h S^{h-\xi} f_* T^\xi u d\xi,$$

$$\|T^h u - T^h v\| \le e^{(\omega+\gamma)h}\|u-v\|, \ \forall h \ge 0, u, v \in AP(\mathbb{X}).$$

From this,

$$[T^h u](t) = [S^h u](t) + \int_0^h [S^{h-\xi} f_*(T^\xi u)](t) d\xi, \forall t \in \mathbb{R}.$$

Thus

$$\begin{aligned}[T^h u](t) &= S(h)u(t-h) + \int_0^h S(h-\xi)[f_*(T^\xi u](t-h+\xi)d\xi \\ &= S(h)u(t-h) + \int_0^h S(h-\xi) f(t+\xi-h, [T^u](t+\xi-h))d\xi \\ &= S(h)u(t-h) + \int_{t-h}^t S(t-\eta) f(\eta, [T^{\eta-(t-h)}u](\eta) d\eta.\end{aligned}$$

If we denote $[T^{t-s}u](t)$ by $x(t)$, we get

$$x(t) = S(t-s)z + \int_s^t S(t-\xi) f(\xi, x(\xi)) d\xi, \forall t \ge s, \tag{5.82}$$

where $z = u(s)$. Consequently, from the uniqueness of mild solutions of Equation (5.78) we get $[T^{t-s}u](t) = x(t) = U(t,s)u(s)$ and $[T^h u](t) = U(t,t-h)u(t-h)$ for all $t \ge s, u \in AP(Q)$. This completes the proof of the proposition. □

5.2.2 *Almost periodic solutions*

5.2.2.1 *Almost periodic solutions of differential equations without delay*

The main idea underlying our approach is the following assertion.

Corollary 5.1. *Let all assumptions of Proposition 5.1 be satisfied. Then a mild solution $x(t)$ of Eq. (5.77), defined on the whole real line $\mathbb{R}$, is almost periodic if and only if it is a common fixed point of the evolution semigroup $(T^h)_{h\geq 0}$ defined in Proposition 5.1.*

Proof. Suppose that $x(t)$, defined on the real line $\mathbb{R}$, is an almost periodic mild solution of Eq. (5.78). Then from the uniqueness of mild solutions we get

$$x(t) = U(t, t-h)x(t-h) = [T^h x](t), \ \forall t \in \mathbb{R}.$$

This shows that x is a fixed point of T^h for every $h > 0$. Conversely, suppose that $y(\cdot)$ is any common fixed point of $T^h, h \geq 0$. Then

$$y(t) = [T^{t-s}y](t) = U(t,s)y(s), \ \forall t \geq s.$$

This shows that $y(\cdot)$ is a mild solution of Eq. (5.78). □

We now apply Corollary 5.1 to find sufficient conditions for the existence of almost periodic mild solutions of Eq. (5.78).

Corollary 5.2. *Let all conditions of Proposition 5.1 be satisfied. Furthermore, let $\omega + \mu$ be negative and $-f_* - \mu I$ be accretive. Then there exists a unique almost periodic mild solution of Eq. (5.78).*

Proof. It is obvious that there exists a unique common fixed point of the semigroup $(T^h)_{h\geq 0}$. The assertion now follows from Corollary 5.1. □

Remark 5.2.

(1) A particular case in which we can check the accretiveness of $-f_* - \mu I$ is $\omega + \gamma < 0$. In fact, this follows easily from the above estimates for $\|u - v\|$ (see the estimate (5.81)).
(2) It is interesting to "compute" the infinitesimal generator of the evolution semigroup $(T^h)_{h\geq 0}$ determined by Proposition 5.1. To this purpose, let us recall the operator L which relates a mild solution u of the equation $\dot{x} = Ax + f(t)$ to the forcing term f by the rule $Lu = f$ (see

Sections 2.1 and 2.2 for more discussion on this operator). From the proof of Proposition 5.1 it follows in particular that *the infinitesimal generator $\mathcal{G}$ of the evolution semigroup $(T^h)_{h\geq 0}$ is $-L + f_*$.*

(3) It can be seen that u is a mild solution of Eq. (5.78) if and only if $(-L + f_*)u = 0$, so it is the fixed point of the semigroup $(T^h)_{h\geq 0}$.

(4) Let f_* act on the function space $\Lambda(\mathbb{X}) \cap AP(\mathbb{X})$. Then by the same argument as in the proof of Proposition 5.1 we can prove that the evolution semigroup $(T^h)_{h\geq 0}$ leaves $\Lambda(\mathbb{X}) \cap AP(\mathbb{X})$ invariant. This will be helpful if we want to discuss the spectrum of the unique almost periodic solution in Corollary 5.2.

5.2.2.2 *Almost periodic solutions of differential equations with delays*

In this subsection we apply the results of the previous subsection to study the existence of almost periodic mild solutions of the equation

$$\frac{dx}{dt} = Ax + f(t, x, x_t) \tag{5.77}$$

where A is defined as in the previous subsection, and f is an everywhere defined continuous mapping from $\mathbb{R} \times \mathbb{X} \times C$ to $\mathbb{X}$. Hereafter we call a continuous function $x(t)$ defined on the real line $\mathbb{R}$ a mild solution of Eq. (5.77) if

$$x(t) = S(t-s)x(s) + \int_s^t S(t-\xi)f(\xi, x(\xi), x_\xi)d\xi, \ \forall t \geq s.$$

We should emphasize that our study is concerned only with the existence of almost periodic mild solutions of Eq. (5.77), and not with all mild solutions in general.

Definition 5.12. (condition H6). Equation (5.77) is said to satisfy *condition H6* if the following is true:

(1) For every $g \in AP(\mathbb{X})$ the mapping $F(t,x) = f(t,x,g_t)$ satisfies conditions $H4$ and $H5$ with the same constant γ.

(2) There exists a constant μ with $\omega - \mu < 0$ such that $-(\mu I + F_*)$ is accretive for every $g \in AP(\mathbb{X})$.

(3) $[x - y, f(t,x,\phi) - f(t,y,\phi')] \leq \gamma\|x-y\| + \delta\|\phi - \phi'\|$, $\forall t \in \mathbb{R}, x, y \in \mathbb{X}, \phi, \phi' \in C$.

Theorem 5.25. *Let condition H6 hold. Then for δ sufficiently small (see the estimate (5.86) below), Eq. (5.77) has an almost periodic mild solution.*

Proof. First we fix a function $g \in AP(\mathbb{X})$. In view of Proposition 5.1 we observe that the equation

$$\frac{dx}{dt} = Ax + F(t,x)$$

has a unique almost periodic mild solution, where $F(t,x) = f(t,x,g_t)$. We denote this solution by Tg. Thus, we have defined an operator T acting on $AP(\mathbb{X})$. We now prove that T is a strict contraction mapping. In fact, let us denote by $U(t,s)$ and $V(t,s)$ the Cauchy operators

$$U(t,s)x = S(t-s)x + \int_s^t S(t-\xi)f(\xi, U(\xi,s)x, g_\xi)d\xi, \tag{5.83}$$

$$V(t,s)x = S(t-s)x + \int_s^t S(t-\xi)f(\xi, V(\xi,s)x, h_\xi)d\xi, \tag{5.84}$$

for given $g, h \in AP(\mathbb{X}), x \in \mathbb{X}, t \geq s$.

Putting $u(t) = U(t,s)x, v(t) = V(t,s)x$ for given s, x, from the assumptions we have

$$[u(t) - v(t), f(t,u(t),g_t - f(t,v(t),h_t] \leq m(t, \|u(t) - v(t)\|),$$

where $m(t, \|u(t) - v(t)\|) = \gamma\|u(t) - v(t)\| + \delta\|h - g\|$. Using this we get

$$\begin{aligned}\|u(t) - v(t)\| \leq \|u(t-\eta) - v(t-\eta)\| &+ \eta m(t, \|u(t) - v(t)\|) \\ &+ \int_{t-\eta}^t \|S(t-\xi)f(\xi,u(\xi),h_\xi) - f(t,u(t),h_t)\|d\xi \\ &+ \int_{t-\eta}^t \|S(t-\xi)f(\xi,v(\xi),g_\xi) - f(t,v(t),g_t)\|d\xi.\end{aligned}$$

Now let us fix arbitrary real numbers $a \leq b$. Since the functions $S(t-\xi)f(\xi,u(\xi),h_\xi)$ and $S(t-\xi)f(\xi,v(\xi),g_\xi)$ are uniformly continuous on the set $a \leq \xi \leq t \leq b$, for every $\varepsilon > 0$ there exists an $\eta_0 = \eta_0(\varepsilon)$ such that

$$\begin{aligned}\|S(t-\xi)f(\xi,u(\xi),h_\xi) - f(t,u(t),h_t)\| < \varepsilon, \\ \|S(t-\xi)f(\xi,v(\xi),g_\xi) - f(t,v(t),g_t)\| < \varepsilon,\end{aligned}$$

for all $\|t - \xi\| < \eta_0$ and $t \leq \xi \in [a,b]$. Hence, denoting $\|u(t) - v(t)\|$ by $\alpha(t)$, for $\eta < \eta_0$ we have

$$\alpha(t) - e^{\omega\eta}\alpha(t-\eta) \leq \eta m(t, \alpha(t)) + 2\eta\varepsilon. \tag{5.85}$$

Applying this estimate repeatedly, we get

$$\alpha(t) - e^{\omega(t-s)} \leq \sum_{i=1}^n e^{\omega(t-t_i)} m(t_i, \alpha(t_i))\Delta_i + 2\varepsilon \sum_{i=1}^n e^{\omega(t-t_i)}\Delta_i,$$

where $t_0 = s < t_1 < t_2 < ... < t_n = t$ and $|t_i - t_{i-1}| = \Delta_i$. Thus, since ε is arbitrary, and since the function m is continuous, we get

$$\begin{aligned}\alpha(t) - e^{\omega(t-s)}\alpha(s) &\leq \int_s^t e^{\omega(t-\xi)} m(\xi, \alpha(\xi)) d\xi \\ &= \int_s^t e^{\omega(t-\xi)} (\gamma\alpha(\xi) + \delta\|h - g\|) d\xi.\end{aligned}$$

Applying Gronwall's inequality we get

$$\alpha(t) \leq e^{(\gamma+\omega)(t-s)}\alpha(s) + e^{\gamma(t-s)+\omega t}\left(\frac{e^{-\omega s} - e^{-\omega t}}{\omega}\right)\delta\|h - g\|.$$

Because of the identity $\alpha(s) = \|u(s) - v(s)\| = \|U(s,s)x - V(s,s)x\| = 0$, from the above estimate we obtain

$$\sup_{t-1\leq\xi\leq t} \|U(\xi, t-1)x - V(\xi, t-1)x\| \leq \frac{e^{\gamma+\omega} - e^{\gamma}}{\omega}\delta\|h - g\|.$$

Now let us denote by $T_h^t, T_g^t, t \geq 0$ the respective evolution semigroups corresponding to Eq. (5.83) and Eq. (5.84). Since Th and Tg are defined as the unique fixed points u_0, v_0 of T_h^1, T_g^1, respectively, we have

$$\begin{aligned}\|Th - Tg\| = \|u_0 - v_0\| &= \|T_h^1 u_0 - T_g^1 v_0\| \leq \\ &\leq \|T_h^1 u_0 - T_g^1 u_0\| + \|T_g^1 u_0 - T_g^1 - v_0\| \\ &\leq \frac{e^{\gamma+\omega} - e^{\gamma}}{\omega}\delta\|h - g\| + e^{\omega-\mu}\|u_0 - v_0\| \\ &= N\delta\|h - g\| + e^{\omega-\mu}\|Th - Tg\|,\end{aligned}$$

where $N = (e^{\gamma+\omega} - e^{\gamma})/\omega$. Finally, we have

$$\|Th - Tg\| \leq \frac{e^{\gamma}(e^{\omega} - 1)}{\omega(1 - e^{\omega-\mu})}.$$

Thus, if the estimate

$$\delta < \frac{\omega(1 - e^{\omega-\mu})}{e^{\gamma}(e^{\omega} - 1)} \tag{5.86}$$

holds true, then T is a strict contraction mapping in $AP(\mathbb{X})$. By virtue of the Contraction Mapping Principle T has a unique fixed point. It is easy to see that this fixed point is an almost periodic mild solution of Eq. (5.77). This completes the proof of the theorem. □

Remark 5.3.

(1) In case $\omega = 0, \gamma = -\mu$ we get the estimate

$$\delta < e^{\mu} - 1 = \mu + \mu^2/2 + ...$$

which guarantees the existence of the fixed point of T.

(2) If $\omega + \gamma < 0$, then we can choose $\mu = -\gamma$, and therefore we get the accretiveness condition on $-(F_* + \mu I)$. However, in general, the condition $\omega + \gamma < 0$ is a very strong restriction on the coefficients of Eq. (5.77), if f depends explicitly on t.

5.2.2.3 *Examples*

In applications one frequently encounters functions f from $\mathbb{R} \times \mathbb{X} \times C \to \mathbb{X}$ of the form

$$f(t, x, g_t) = F(t, x) + G(t, g_t), \ \forall t \in \mathbb{R}, \ x \in \mathbb{X}, \ g_t \in C,$$

where F satisfies condition ii) of Definition 5.12 and $G(t, y)$ is Lipschitz continuous with respect to $y \in C$, i.e.

$$\|G(t, y) - G(t, z)\| \leq \delta\|y - z\|, \ \forall t \in \mathbb{R}, \ y, z \in C$$

for some positive constant δ. In order to describe a concrete example we consider a bounded domain Ω in $\mathbb{R}^n$ with smooth boundary $\partial\Omega$ and suppose that

$$A(x, D)u = \sum_{|\alpha| \leq 2m} a_\alpha(x) D^\alpha u$$

is a strongly elliptic differential operator in Ω. Then, defining the operator

$$Au = A(x, D)u, \ \forall u \in D(A) = W^{2m,2}(\Omega) \cap W_0^{m,2}(\Omega)$$

we know from Theorem 3.6 in [Pazy (90)] that the operator $-A$ is the infinitesimal generator of an analytic semigroup of contractions on $L^2(\Omega)$. Now let $f, g : \mathbb{R} \times \Omega \times \mathbb{R} \to \mathbb{R}$ be Lipschitz continuous and define the operators $F(t, w)(x) = f(t, x, w(x))$ and $G(t, w)(x) = g(t, x, w(x))$ where $t \in \mathbb{R}$, $x \in \Omega$ and $w \in L^2(\Omega)$. Then, for any positive constant r, the boundary value problem

$$\frac{\partial u(t, x)}{\partial t} = A(x, D)u(t, x) + f(t, x, u(t, x)) + g(t, x, u(t - r, x)) \quad \text{in } \Omega \ ,$$

$$u(t, x) = 0 \quad \text{on } \partial\Omega$$

fits into the abstract setting of Eq. (5.77).

5.3 Comments and Further Reading Guide

5.3.1 *Further Reading Guide*

For Eq. (5.22), the idea of deriving periodic solutions using boundedness has been recently extended to general fading memory phase spaces with axioms in [Ezzinbi, Liu and Minh (32)]. So that in some sense, the study along this line is getting complete.

A phase space for Eq. (5.22) is called a *fading memory space* if it is a Banach space $(\Gamma, \|\cdot\|_\Gamma)$ consisting of functions from $(-\infty, 0]$ to $\mathbb{X}$ that satisfy the following axioms ([Hale and Kato (41); Hino and Murakami (47)]):

(A1). There exist a positive constant H and locally bounded non-negative continuous functions $K(\cdot)$ and $M(\cdot)$ on $[0, \infty)$ with the property that if $u : (-\infty, a) \to X$ is continuous on $[\sigma, a)$ with $u_\sigma \in \Gamma$ for some $\sigma < a$, then for all $t \in [\sigma, a)$,

  (i). $u_t \in \Gamma$,

  (ii). u_t is continuous in t (with respect to $\|\cdot\|_\Gamma$),

  (iii). $H\|x(t)\| \leq \|x_t\|_\Gamma \leq K(t-\sigma)\sup_{\sigma \leq s \leq t}\|x(s)\| + M(t-\sigma)\|x_\sigma\|_\Gamma$.

(A2). If $\{\phi^k\}$, $\phi^k \in \Gamma$, converges to ϕ uniformly on any compact set in $(-\infty, 0]$ and if $\{\phi^k\}$ is a Cauchy sequence in Γ, then $\phi \in B$ and $\phi^k \to \phi$ in Γ, $k \to \infty$.

A fading memory space is called a *uniform fading memory space* if it satisfies (A1) and (A2) with $K(\cdot) \equiv K_1$ (a constant) and $M(t) \to 0$ as $t \to \infty$.

In [Ezzinbi, Liu and Minh (32)], the Poincaré operator is shown to be condensing in Γ under the condition that

$$M(0) < 1,$$

which simplifies a condition in [Henriquez (45)] of the form

$$\inf_{0<\sigma<T} M(T-\sigma)[\frac{1}{H}K(\sigma)\sup_{0\leq t\leq\sigma}\|T(t)\| + M(\sigma)] < 1. \tag{5.87}$$

Then, when the phase space Γ is a uniform fading memory space, it is shown that Eq.(5.22) has periodic solutions if its solutions are ultimate bounded.

5.3.2 *Comments*

Most results here are obtained under the condition that the nonlinear function f is Lipschitzian in variables other than t. If f is an arbitrary nonlinear function, then we need to also assume that f maps a bounded set into a bounded set and require that the solutions are also bounded (in addition to being ultimate boundeded) in order to carry the proofs.

Appendix

A.1 Lipschitz Operators

Let $\mathbf{X}$ and $\mathbf{Y}$ be given Banach spaces over the same field $\mathbf{R}$, $\mathbf{C}$. An operator $A : \mathbf{X} \to \mathbf{Y}$ is called *Lipschitz continuous* if there is a positive constant L such that

$$\|Ax - Ay\|_{\mathbf{Y}} \leq L\|x - y\|_{\mathbf{X}}, \ \forall x, y \in \mathbf{X}.$$

For a Lipschitz continuous operator A the following

$$\|A\| := \sup_{x,y \in \mathbf{X},\ x \neq y} \|Ax - Ay\|/\|x - y\|$$

is finite and is called the *Lipschitz constant* of A. The set of all Lipschitz continuous operators from $\mathbf{X}$ to $\mathbf{Y}$ is denoted by $\mathcal{L}ip(\mathbf{X}; \mathbf{Y})$ and $\mathcal{L}ip(\mathbf{X}; \mathbf{X}) = \mathcal{L}ip(\mathbf{X})$ for short. A member $A \in \mathcal{L}ip(\mathbf{X})$ is said to be *invertible* if there is a $B \in \mathcal{L}ip(\mathbf{X})$ such that $A \cdot B = B \cdot A = I$. B is called the *inverse* of A and is denoted by A^{-1}

Theorem A.26. *(Lipschitz Inverse Mapping) Let $\mathbf{X}$ be a Banach space, A is an invertible member of $\mathcal{L}ip(\mathbf{X})$ and B is a member of $\mathcal{L}ip(\mathbf{X})$ such that $\|B\| \cdot \|A^{-1}\| < 1$. Then $A + B$ is invertible in $\mathcal{L}ip(\mathbf{X})$ and*

$$\|(A + B)^{-1}\| \leq \|A^{-1}\|(1 - \|B\| \cdot \|A^{-1}\|)^{-1}.$$

Proof. We first prove the following assertion: If $A \in \mathcal{L}ip(\mathbf{X})$ such that $\|A\| < 1$. Then $(I - A)$ is invertible in $\mathcal{L}ip(\mathbf{X})$ and

$$\|(I - A)^{-1}\| \leq (1 - \|A\|)^{-1}. \tag{A.1}$$

In fact, for $x, y \in \mathbf{X}$

$$\|(I - A)x - (I - A)y\| \geq \|x - y\| - \|Ax - Ay\| \geq (1 - \|A\|)\|x - y\|.$$

Thus $I - A$ is injective. If $z, w \in R(I - A)$, then

$$\|(I - A)^{-1}z - (I - A)^{-1}w\| \leq (1 - \|A\|)^{-1}\|z - w\|.$$

For $x \in \mathbf{X}$ by induction we can prove that

$$\|B_{n+1}x - B_n x\| \leq \|A\|^n \|Ax\|, \ \forall n = 0, 1, 2, ...$$

where by induction we define $B_0 := I$, $B_n := I + AB_{n-1}$, $\forall n = 1, 2....$ Indeed, this holds true for $n = 0$, so if we assume it to be true for $n - k$, then

$$\begin{aligned}\|B_{k+1}x - B_k x\| &= \|AB_k x - AB_{k-1}x\| \\ &\leq \|A\| \cdot \|B_k x - B_{k-1}x\| \\ &\leq \|A\| \cdot \|A\|^{k-1}\|Ax\|,\end{aligned}$$

so the assertion follows by induction. For any positive integer p,

$$\begin{aligned}\|B_{n+p}x - B_n x\| &= \left\| \sum_{k=0}^{p-1} (B_{n+k+1}x - B_{n+k}x) \right\| \\ &\leq \sum_{k=0}^{p-1} \|(B_{n+k+1}x - B_{n+k}x)\| \\ &\leq \sum_{k=0}^{p-1} \|A\|^{n+k}\|Ax\| \leq \|A^n\|\|Ax\|(1 - \|A\|)^{-1}.\end{aligned}$$

Since $\|A\| < 1$ and $\mathbf{X}$ is a Banach space, $Cx = \lim_{m \to \infty} B_m x$ exists for all $x \in \mathbf{X}$ and

$$\|Cx - B_n x\| = \lim_{p \to \infty} \|B_{n+p}x - B_n x\| \leq \|A\|^n \|Ax\|(1 - \|A\|)^{-1}.$$

Since A is continuous,

$$Cx = \lim_{n \to \infty} B_n x = \lim_{n \to \infty} (I - AB_{n-1})x = x + ACx.$$

This shows that $C = I + AC$, so C is a right inverse of $I - A$, i.e. $(I - A)C = I$. Finally, this shows the surjectiveness of $I - A$, proving the assertion that $I - A$ is invertible in $\mathcal{L}ip(\mathbf{X})$.

We are now in a position to prove the theorem. In fact, we have $(A + B) = (I + BA^{-1})A$ and $\|BA^{-1}\| \leq \|B\| \cdot \|A^{-1}\| < 1$. By the above assertion, $(I + BA^{-1})^{-1}$ exists as an element of $\mathcal{L}ip(\mathbf{X})$. Hence $(A + B)^{-1} = A^{-1}(I + BA^{-1})^{-1}$. and

$$\|(I + BA^{-1})^{-1}\| \leq (1 - \|B\| \cdot \|A^{-1}\|)^{-1}.$$

□

A modification of the above theory for Lipschitz continuous operators from a Banach space $\mathbf{X}$ to another Banach space $\mathbf{Y}$ can be easily made. For instance, the following is true:

Theorem A.27. *Let A be an invertible member of $\mathcal{L}ip(\mathbf{X}, \mathbf{Y})$. Then for sufficiently small positive k, the operator $A + B$ is an invertible member of $\mathcal{L}ip(\mathbf{X}, \mathbf{Y})$ if $\|B\| < k$.*

Proof. Set $C = A^{-1}(A + B) - I$. Then C is a member of $\mathcal{L}ip(\mathbf{X})$, and for all $x, y \in \mathbf{X}$,

$$\begin{aligned}\|Cx - Cy\|_{\mathbf{X}} &= \|(A^{-1}(A + B) - A^{-1}A)x - (A^{-1}(A + B) - A^{-1}A)y\| \\ &\leq \|A^{-1}\| \cdot \|B\| \cdot \|x - y\|.\end{aligned}$$

Thus for sufficiently small positive k, $I + C$ is invertible, so is $A + B$. □

A.2 Fixed Point Theorems

The following fixed point theorems can be found in Smart [Smart (99)], Burton [Burton (18)], and Hale and Lunel [Hale and Lunel (42)].

Definition. *A mapping (operator) P on a metric space (X, ρ) is called a* **contraction mapping** *if there is an $r \in (0, 1)$ such that*

$$\rho(Px, Py) \leq r\rho(x, y).$$

Theorem (Contraction mapping principle). *Let P be a contraction mapping on a complete metric space X, then there is a unique $x \in X$ with $Px = x$. Moreover, $x = \lim_{n\to\infty} x_n$, where x_0 is any element of X and $x_{j+1} = Px_j$, $j = 0, 1, \cdots$.*

Proof. Now, for some $0 < r < 1$, we have $\rho(Py, Pz) \leq r\rho(y, z)$ when $y, z \in X$. Let x_0 be any element of X and define $x_{j+1} = Px_j$, $j = 0, 1, \cdots$. Then $x_1 = Px_0$, $x_2 = Px_1 = P^2x_0, \cdots, x_j = Px_{j-1} = \cdots = P^jx_0$, $j =$

$1, 2, \cdots$. Thus, for $m > n$,

$$\begin{aligned}
\rho(x_n, x_m) &= \rho(P^n x_0, P^m x_0) \\
&\le r\rho(P^{n-1}x_0, P^{m-1}x_0) \\
&\vdots \\
&\le r^n \rho(x_0, P^{m-n}x_0) = r^n \rho(x_0, x_{m-n}) \\
&\le r^n \Big[\rho(x_0, x_1) + \rho(x_1, x_2) + \cdots + \rho(x_{m-n-1}, x_{m-n})\Big] \\
&\le r^n \Big[\rho(x_0, x_1) + r\rho(x_0, x_1) + \cdots + r^{m-n-1}\rho(x_0, x_1)\Big] \\
&= r^n \rho(x_0, x_1)\Big[1 + r + \cdots + r^{m-n-1}\Big] \\
&\le r^n \rho(x_0, x_1)\frac{1}{1-r}. \qquad \text{(A.2)}
\end{aligned}$$

As $0 < r < 1$, the right-hand side goes to zero when $n \to \infty$. Thus $\{x_n\}$ is a Cauchy sequence, and hence has a limit $x \in X$ because X is a complete metric space. Now, it is easily seen that P is continuous, therefore

$$Px = P\Big(\lim_{n\to\infty} x_n\Big) = \lim_{n\to\infty}\Big(Px_n\Big) = \lim_{n\to\infty} x_{n+1} = x, \qquad \text{(A.3)}$$

and x is a fixed point of P. If y is also a fixed point of P, then

$$\rho(x, y) = \rho(Px, Py) \le r\rho(x, y), \qquad \text{(A.4)}$$

and, as $0 < r < 1$, we must have $\rho(x, y) = 0$, which implies $x = y$. This completes the proof.

Theorem (Brouwer's fixed point theorem). *Let $B \subset \Re^n$ be nonempty, convex, and compact, and let $F : B \to B$ be a continuous operator. Then F has a fixed point in B.*

Theorem (Schauder's first fixed point theorem). *Let X be a nonempty, convex, and compact subset of a Banach space Y, and let $P : X \to X$ be a continuous operator. Then P has a fixed point in X.*

Theorem (Schauder's second fixed point theorem). *Let X be a nonempty, convex, and bounded subset of a Banach space Y and let $P : X \to X$ be a compact operator. Then P has a fixed point in X.*

The following are called "asymptotic fixed point theorems" since they use the idea that if P^m has a unique fixed point for some positive integer m, then P itself has a fixed point.

Definition. *Let A and B be subsets of a Banach space Z. If $A = B \cap C$ for an open subset C of Z, then A is open relative to B.*

Theorem (Horn's fixed point theorem). *Let $E_0 \subset E_1 \subset E_2$ be convex subsets of a Banach space Z, with E_0 and E_2 compact subsets and E_1 open relative to E_2. Let $P : E_2 \to Z$ be a continuous operator such that for some integer m, one has*

$$P^j(E_1) \subset E_2, \quad 1 \leq j \leq m-1, \tag{A.5}$$

$$P^j(E_1) \subset E_0, \quad m \leq j \leq 2m-1, \tag{A.6}$$

then P has a fixed point in E_2.

Theorem (Browder's fixed point theorem). *Let $E_0 \subset E_1 \subset E_2$ be convex subsets of a Banach space Z, with E_0 closed and E_1, E_2 open. Let $P : E_2 \to Z$ be a compact operator such that for some integer m, one has*

$$P^j(E_0) \subset E_1, \quad 0 \leq j \leq m, \tag{A.7}$$

$$P^m(E_1) \subset E_0, \tag{A.8}$$

then P has a fixed point in E_2.

Theorem (Hale and Lunel's fixed point theorem). *Suppose $S_0 \subseteq S_1 \subseteq S_2$ are convex bounded subsets of a Banach space Y, S_0 and S_2 are closed, and S_1 is open in S_2, and suppose $P : S_2 \to Y$ is (S_2)-condensing in the following sense: if U and $P(U)$ are contained in S_2 and $\alpha(U) > 0$, then $\alpha(P(U)) < \alpha(U)$. If $P^j(S_1) \subseteq S_2$, $j \geq 0$, and, for any compact set $H \subseteq S_1$, there is a number $N(H)$ such that $P^k(H) \subseteq S_0$, $k \geq N(H)$, then P has a fixed point.*

A.3 Invariant Subspaces

Let $S \subset H$ be a closed subset and P_S, the orthogonal projection onto the subspace S. The operator is still a densely defined closed (possibly unbounded) linear operator in H.

Definition A.13. S is said to be an *invariant subspace* for A if we have the inclusion $A(D(A) \cap S) \subset S$.

Example A.2. Let us mention the following classical invariant subspaces for the closed unbounded linear operator A defined into the Hilbert space H.

1. $S = N(A) = \{x \in D(A) : \; Ax = 0\}$ is an invariant subspace for A.

2. If A is a self-adjoint linear operator, then any eigenspace $S_\lambda = N(\lambda I - A)$ is an invariant for A. In fact it can be easily shown that S_λ reduces A.

Theorem A.28. *The equality $P_S A P_S = A P_S$ is a necessary and sufficient condition for a subspace S to be invariant for a linear operator A.*

Proof. Assume $P_S A P_S = A P_S$ and if $x \in D(A) \cap S$, then $x = P_S x \in D(A)$ and $Ax = AP_S x = P_S A P_S x \in S$.

Conversely, if S is invariant for A; let $x \in H$ be such that $P_S x \in D(A)$. Then $AP_S x \in S$ and then $P_S A P_S x = A P_S x$. Therefore $AP_S \subset P_S A P_S$. Since $D(AP_S) = D(P_S A P_S)$, it turns out that $AP_S = P_S A P_S$. □

Definition A.14. A closed proper subspace S of the Hilbert space H is said to reduce an operator A if $P_S D(A) \subset D(A)$ and both S and $H \ominus S$, the orthogonal complement of S, are invariant for A.

Using the above Theorem , the following key result can be proved.

Theorem A.29. *A closed subspace S of H reduces an operator A if and only if $P_S A \subset A P_S$.*

Proof. See the proof in [Locker (64)] *Theorem 4.11., p. 29.* □

Remark A.4. In fact the meaning of the inclusion $P_S A \subset AP_S$ is that: if $x \in D(A)$, then $P_S x \in D(A)$ and $P_S A x = A P_S x$.

A.4 Semilinear Evolution Equations

We recall in this section a result on the well posedness for semilinear equations of the form

$$\frac{dx}{dt} = Ax + Bx \; , \quad x \in \mathbf{X} \tag{A.9}$$

where $\mathbf{X}$ is a Banach space, A is the infinitesimal generator of a C_0-semigroup $S(t)$, $t \geq 0$ of linear operators of type ω, i.e.

$$\|S(t)x - S(t)y\| \leq e^{\omega t}\|x - y\|, \; \forall \, t \geq 0, \; x, y \in \mathbf{X} \, ,$$

and B is an everywhere defined continuous operator from $\mathbf{X}$ to $\mathbf{X}$. Hereafter, by a mild solution $x(t), t \in [s, \tau]$ of equation (A.9) we mean a continuous solution of the integral equation

$$x(t) = S(t-s)x + \int_s^t S(t-\xi)Bx(\xi)d\xi, \; \forall s \leq t \leq \tau. \tag{A.10}$$

Before proceeding we recall some notions and results which will be frequently used later on. We define the bracket $[\cdot,\cdot]$ in a Banach space $\mathbf{Y}$ as follows (see e.g. [Martin (67)] for more information)

$$[x,y] = \lim_{h\to+0} \frac{\|x+hy\|-\|y\|}{h} = \inf_{h>0} \frac{\|x+hy\|-\|y\|}{h}$$

Definition A.15. Suppose that F is a given operator on a Banach space $\mathbf{Y}$. Then $(F+\gamma I)$ is said to be *accretive* if and only if for every $\lambda > 0$ one of the following equivalent conditions is satisfied

(1) $(1-\lambda\gamma)\|x-y\| \leq \|x-y+\lambda(Fx-Fy)\|, \ \forall x,y \in D(F)$,
(2) $[x-y, Fx-Fy] \geq -\gamma\|x-y\|, \ \forall x,y \in D(F)$.

In particular, if $\gamma = 0$, then F is said to be accretive.

Remark A.5. From this definition we may conclude that $(F+\gamma I)$ is accretive if and only if

$$\|x-y\| \leq \|x-y+\lambda(Fx-Fy)\| + \lambda\gamma\|x-y\| \qquad \text{(A.11)}$$

for all $x,y \in D(F), \lambda > 0, 1 \geq \lambda\gamma$.

Theorem A.30. *Let the above conditions hold true. Then for every fixed* $s \in \mathbf{R}$ *and* $x \in \mathbf{X}$ *there exists a unique mild solution* $x(\cdot)$ *of Eq.(A.9) defined on* $[s,+\infty)$. *Moreover, the mild solutions of Eq.(A.9) give rise to a semigroup of nonlinear operators* $T(t), t \geq 0$ *having the following properties:*

$$i) \ T(t)x = S(t)x + \int_0^t S(t-\xi)BT(\xi)xd\xi, \ \forall t \geq 0, x \in \mathbf{X}, \quad \text{(A.12)}$$

$$ii) \ \|T(t)x - T(t)y\| \leq e^{(\omega+\gamma)t}\|x-y\|, \ \forall t \geq 0, x,y \in \mathbf{X}. \quad \text{(A.13)}$$

More detailedly information on this subject can be found in [Martin (67)].

Theorem A.31. *Let* D *be a closed and convex subset of a Hausdorff locally convex space such that* $0 \in D$, *and let* G *be a continuous mapping of* D *into itself. If the implication*

$$(V = \text{conv}G(V) \quad or \quad V = G(V) \cup \{0\}) \Longrightarrow V \textit{ is relatively compact}$$

holds for every subset V *of* D, *then* G *has a fixed point.*

In what follows we shall need the following definition of the proper mapping at a given point.

Let X, Y be metric spaces, $f : X \to Y$ be a continuous function and let $y \in Y$. f is said to be proper at the point y provided that there exists $\varepsilon > 0$ such that for any compact set $K \subset B(y, \varepsilon)$ the set $f^{-1}(K)$ is compact, where $B(y, \varepsilon)$ is the open ball in Y of center y and radius ε.
If $f^{-1}(K)$ is compact for any compact $K \subset Y$, then f is called proper.

The proofs of Aronszajn type results (see [Aronszjan (9)]) are based on the following Browder-Gupta type theorem (see [Górniewicz (38)])

Theorem A.32. *Let E be a Banach space and $f : X \to E$ be a continuous map such that the following conditions are satisfied:*

(i) f is proper at $0 \in E$,
(ii) for every $\varepsilon > 0$ there exists a continuous map $f_\varepsilon : X \to E$ for which we have:

(a) $\|f(x) - f_\varepsilon(x)\| < \varepsilon$ for every $x \in X$,
(b) the map $\tilde{f}_\varepsilon : f_\varepsilon^{-1}(B(0,\varepsilon)) \to B(0,\varepsilon)$, $\tilde{f}_\varepsilon(x) = f_\varepsilon(x)$ for every $x \in f_\varepsilon^{-1}(B(0,\varepsilon))$, is a homeomorphism.

Then the set $f^{-1}(\{0\})$ is an R_δ set.

Bibliography

H. Amann, "Ordinary Differential Equations", de Gruyter Studies in Math. **13**, Walter de Gruyter, Berlin 1990.
H. Amann, *Periodic solutions of semi-linear parabolic equations*, Nonlinear Analysis, A Collection of Papers in Honor of Erich Roth, Academic Press, New York, 1978, 1-29.
L. Amerio, G. Prouse, "Almost Periodic Functions and Functional Equations", Van Nostrand Reinhold, New York, 1971.
W. Arendt and C.J.K. Batty, Tauberian theorems for one-parameter semigroups, *Trans. Am. Math. Soc.* **306** (1988), 837-852.
W. Arendt, C.J.K. Batty, Almost periodic solutions of first and second oder Cauchy problems, *J. Diff. Eq* **137**(1997), N.2, 363-383.
W. Arendt, F. Räbiger, A. Sourour, Spectral properties of the operators equations AX + XB = Y, *Quart. J. Math. Oxford (2)*, **45**(1994), 133-149.
W. Arendt, C. J. K. Batty, M. Hieber, F. Neubrander, "Vector-valued Laplace transforms and Cauchy problems", Monogrphs in mathematics. *Birkhauser Verlag, Basel-Boston-Berlin*, **96**(2001).
W. Arendt, S. Bu, *The operator-valued Marcinkiewicz multiplier theorem and maximal regularity,* Math. Z. **240** (2002), 311-343.
N. Aronszjan, *Le correspondant topologique de l'unicité dans la theorie des équations différentielles*, Ann. Math. **43** (1942), 730-738.
B. Aulbach , N.V. Minh, Nonlinear semigroups and the existence, stability of semilinear nonautonomous evolution equations, *Abstract and Applied Analysis* **1**(1996), 351-380.
B. Aulbach, N.V. Minh, Almost periodic mild solutions of a class of partial functional differential equations, *Abstract and Applied Analysis.* To appear.
B. Basit, Harmonic analysis and asymptotic behavior of solutions to the abstract Cauchy problem, *Semigroup Forum* **54**(1997), 58-74.
A.G. Baskakov, Semigroups of difference operators in the spectral analysis of linear differential operators, *Funct. Anal. Appl.* **30**(1996), no. 3, 149–157 (1997).
C.J.K. Batty, W. Hutter, F. Räbiger, Almost periodicity of mild solutions of inhomogeneous periodic Cauchy problems, *J. Diff. Eq.*, **156** (1999), 309-327.
C. J. K. Batty, R. Chill, Y. Tomilov, Strong stability of bounded evolution

families and semigroups. *J. Funct. Anal.* **193** (2002), 116–139.
D. Bugajewska, *On the structure of solution sets of diffential equations in Banach spaces* Math. Slovaca, **50** (4) (2000), 463-471.
D. Bugajewski, G. M. N'Gurkata, On the topological structure of almost automorphic and asymptotically almost automorphic solutions of differential and integral equations in abstract spaces, *Nonlinear Analysis*, **59** (2004), 1333-1345.
T. Burton, "Stability and Periodic Solutions of Ordinary and Functional Differential Equations", Academic Press, Orlando, Florida. 1985.
C. Chicone, Y. Latushkin, *"Evolution semigroups in dynamical systems and differential equations"*. Mathematical Surveys and Monographs, 70. American Mathematical Society, Providence, RI, 1999.
S.N. Chow, J.K. Hale, Strongly limit-compact maps, *Funkc. Ekvac* **17**(1974), 31-38.
W.A. Coppel, "Dichotomies in Stability Theory", Lecture Notes in Math. vol. 629, Springer- Verlag, Berlin - New York, 1978.
C.M. Dafermos, Almost periodic processes and almost periodic solutions of evolution equations, *in* "Dynamical Systems, Proceedings of a University of Florida International Symposium, 1977"Academic Press, pp. 43-57.
G. Da Prato, P. Grisvard, Sommes d'operateurs lineares et equations differentielles operationelles, *J. Math. Pures Appl.* **54** (1975), 305-387.
Ju. L. Daleckii and M.G. Krein, "Stability of Solutions of Differential Equations in Banach Space", Amer. Math. Soc., Providence, RI, 1974.
R. Datko, Uniform asymptotic stability of evolutionary processes in a Banach space. *SIAM J. Math. Anal.* **3** (1972), 428–445.
D. Daners, P.K. Medina, "Abstract Evolution Equations, Periodic Problems and Applications", Pitman Research Notes in Math. Ser. volume 279, Longman. New York 1992.
E.B. Davies, "One-parameter Semigroups", Academic Press, London, 1980.
T. Diagana, G. M. N'Guérékata and Nguyen Van Minh, "Almost Automorphic Solutions of Evolution Equations", Proc. Amerc. Math. Soc. **132** (2004), 3289-3298.
N. Dunford and J.T. Schwartz, "Linear Operators, Part 1", Wiley-Interscience, New York, 1988.
K.J. Engel, R. Nagel, "One-parameter Semigroups for linear Evolution Equations". Springer, Berlin, 1999.
K. Ezzinbi and J. Liu, *Periodic solutions of nondensely defined delay evolution equations*, Journal of Applied Mathematics and Stochastic Analysis, **15**(2002), 113-123.
K. Ezzinbi, J. Liu and N. Minh, Periodic solutions in fading memory spaces. Proceedings of the fifth international conference on dynamical systems and differential equations, *Discrete and Continuous Dynamical Systems.* (2005), suppl., 250–257.
M. Farkas, "Periodic motions", Applied Mathematical Sciences, 104. Springer-Verlag, New York, 1994.
A.M. Fink, "Almost Periodic Differential Equations", Lecture Notes in Math., **377**, Springer Verlag, Berlin - New York, 1974.

T. Furumochi, T. Naito, Nguyen Van Minh, Boundedness and almost periodicity of solutions of partial functional differential equations, *J. Differential Equations*, **180** (2002), 125-152.

J.A. Goldstein, "Semigroups of Linear Operators and Applications", Oxford Mathematical Monographs, Oxford University Press, Oxford 1985.

J. A. Goldstein and G. M. N'Guérékata, "Almost Automorphic Solutions of Semilinear Evolution Equations", Proc. Amer. Math. Soc. (to appear).

L. Górniewicz, *Topological structure of solution sets: current results*, Arch. Math. (Born) **36** (2000), 343-382.

J.K. Hale, *"Theory of Functional Differential Equations"*, Springer-Verlag, New York - Berlin 1977.

J. Hale, *"Asymptotic behavior of dissipative systems"*, American Math. Soc., Providence, Rhode Island, 1988.

J. Hale and J. Kato, *Phase space for retarded equations with infinite delay*, Funkcial. Ekvac., **21**(1978), 11-41.

J. Hale and S. Lunel, *Introduction to Functional Differential Equations*, Springer-Verlag, New York, 1993, 113-119.

A. Haraux, "Nonlinear evolution equations - global behavior of solutions". Lecture Notes in Mathematics, **841**.Springer-Verlag, Berlin - Heidelberg - New York, 1981.

L. Hatvani, T. Kristin, On the existence of periodic solutions for linear inhomogeneous and quasilinear functional differential equations, *J. Diff. Eq.* **97**(1992), 1-15.

H. Henriquez, *Periodic solutions of quasi-linear partial functional differential equations with unbounded delay*, Funkcial. Ekvac., **37**(1994), 329-343.

D. Henry, *"Geometric Theory of Semilinear Parabolic Equations"*, Lecture Notes in Math., Springer-Verlag, Berlin-New York, 1981.

Y. Hino and S. Murakami, *Almost automorphic solutions for abstract functional differential equations*, J. Math. Anal. Appl., **286**(2003), 741-752.

Y. Hino, S. Murakami, T. Naito, "Functional Differential Equations with Infinite Delay", Lect. Notes Math. 1473, Springer-Verlag, 1991.

Y. Hino, T. Naito, N.V. Minh, J.S. Shin, "Almost Periodic Solutions of Differential Equations in Banach Spaces". Taylor & Francis, London - New York, 2002.

Y. Katznelson, "An Introduction to Harmonic Analysis", Dover Publications, New York , 1968.

V. Keyantuo, C. Lizama, *Fourier multipliers and integro-differential equations in Banach spaces.* J. London Math. Soc.**69** (3) (2004), 737-750.

V. Lakshmikantham and S. Leela, *Differential and integral inequalities*, Vol. 1, Academic Press, New York, 1969.

C. Langenhop, Periodic and almost periodic solutions of Volterra integral differential equations with infinite memory, *J. Diff. Eq.* **58**(1985),391-403.

R. Larsen, *"Functional Analysis"*, Decker Inc. New York, 1973.

Yu. Latushkin, S. Monthomery-Smith, Evolutionary semigroups and Lyapunov theorems in Banach spaces, *J. Func. Anal.* **127**(1995), 173-197.

Y. Latushkin, Y. Tomilov, Fredholm Properties of Evolution Semigroups, *Jour-*

nal of Differential Equations **208** (2005), 388-429.

Y. Latushkin, F. Räbiger, Operator valued Fourier multipliers and stability of strongly continuous semigroups. *Integral Equations Operator Theory*, **51** (2005), 375-394.

B.M. Levitan, V.V. Zhikov, "Almost Periodic Functions and Differential Equations", Moscow Univ. Publ. House 1978. English translation by Cambridge University Press 1982.

J. Liu, *A First Course in the Qualitative Theory of Differential Equations*, Prentice Hall, New Jersey, 2003.

J. Liu, *Bounded and periodic solutions of semi-linear evolution equations*, Dynamic Sys. & Appl., **4**(1995), 341-350.

J. Liu, *Bounded and periodic solutions of finite delay evolution equations*, Nonlinear Anal., **34**(1998), 101-111.

J. Liu, *Periodic solutions of infinite delay evolution equations*, J. Math. Anal. Appl., **247**(2000), 627-644.

J. Liu, T. Naito, and N. Minh, *Bounded and periodic solutions of infinite delay evolution equations*, J. Math. Anal. Appl., **286**(2003), 705-712.

J. Locker, "Spectral Theory of Non-Self-Adjoint Two-Point Differential Operators", Amer. Math. Soc. Mathematical Surveys and Monographs, Vol. 73, (2000).

A. Lunardi, "Analytic Semigroups and Optimal Regularity in Parabolic Problems", Birhauser, Basel, 1995.

Yu. I. Lyubich, Vu Quoc Phong, Asymptotic stability of linear differential equations on Banach spaces, *Studia Math.* **88** (1988), 37-42.

R. Martin, "Nonlinear operators and differential equations in Banach spaces", Wiley-Interscience, New York 1976.

J.L. Massera, The existence of periodic solutions of systems of differential equations, *Duke Math. J.***17**, (1950). 457–475.

N.V. Minh, On the proof of characterisations of the exponential dichotomy, *Proc. Amer. Math. Soc* **127**(1999), 779-782.

N.V. Minh, F. Räbiger, R. Schnaubelt, On the exponential stability, exponential expansiveness and exponential dichotomy of evolution equations on the half line, *Int. Eq. and Oper. Theorey* **32**(1998), 332-353.

S. Murakami, T. Naito, N.V. Minh, Evolution semigroups and sums of commuting operators: a new approach to the admissibility theory of function spaces, *J. Diff. Eq.*, **164** (2000), 240-285.

Nguyen Van Minh, Ha Binh Minh, A Marssera - type theorem for almost periodic solutions of higher order delay or advance abstract functional differential equations, *Abstract and Applied Analysis* Vol. **2004** (2004), 881-896.

R. Nagel ,(Ed) "One-parameter Semigroups of Positive Operators", Springer. Lec. Notes in Math. 1184(1986).

T. Naito and N.V. Minh, Evolution semigroups and spectral criteria for almost periodic solutions of periodic evolution equations, *J. Diff. Eq.* **152**(1999), 358-376.

T. Naito, N.V. Minh, R. Miyazaki, J.S. Shin, A decomposition theorem for bounded solutions and the existence of periodic solutions of periodic differential equations, *J. Diff. Eq.* **160**(2000), 263-282.

T. Naito, N.V. Minh, J. S. Shin, New spectral criteria for almost periodic solutions of evolution equations, *Studia Mathematica* **145** (2001), 97-111.
T. Naito, N.V. Minh, R. Miyazaki, Y. Hamaya, Boundedness and almost periodicity in dynamical systems. ıJournal of Difference Equations and Applications, **7** (2001), 507-528.
J. van Neerven, "The asymptotic Behaviour of Semigroups of Linear Operator", Birkhaüser Verlag. Basel. Boston. Berlin, Operator Theory, Advances and Applications Vol.88 1996.
Gaston M. N'Guérékata, "Almost Automorphic and Almost Periodic Functions in Abstract Spaces", Kluwer Academic/ Plenum Publishers, New York, 2001.
G. M. N'Guérékata and T. Diagana, Some remarks on almost automorphic solutions of some abstract differential equations, Far. East J. Math. Sci., (FJMS) 8 (3) (2003), 313-322.
G. M. N'Guérékata, An extension of the Bohr-Neugebauer theorem,*Dynamic Syst. and Appl.*, **10** (2001), 451-454.
G. M. N'Guérékata, Almost automorphic solutions of some differential equations in Banach spaces, *Internat. J. Math. & Math. Sci.*, **23**, **(5)**, (2000), 361-365.
G. M. N'Guérékata, Almost automorphic functions and applications to abstract evolution equations, *Contemporary Math.*, **252**,(1999),71-76.
G. M. N'Guérékata, Almost automorphy, almost periodicity and stability of motions in Banach spaces, *Forum Math.*, **13**, (2001), 581-588.
G. M. N'Guérékata, On almost automorphic solutions of linear operational-differential equations, *Internat. J. Math. & Math. Sci. Vol. 2004, Issue 22.*
G. M. N'Guérékata, On weak almost periodic mild solutions of some linear abstract differential equations, *Proceedings of the Fourth International Conference on Dynamical Systems and Differential Equations,May 24-27, 2002, Wilmington, NC, USA*, (2003), 683-673.
G. M. N'Guérékata, Existence and uniqueness of almost automorphic mild solutions to some semilinear abstract differential equations, *Semigroup Forum*, **69**, No. 1 (2004), 80-86.
G. M. N'Guérékata, "Topics in Almost Automorphy", Springer, New York-Boston-Dordrecht-London-Moscow, 2005.
R.D. Nussbaum, The radius of the essential spectrum, *Duke Math. J.*, **37**(1970), 473-478.
A. Pazy, "Semigroups of Linear Operators and Applications to Partial Differential Equations", Applied Math. Sci. 44, Spriger-Verlag, Berlin-New York 1983.
J. Pruss, "Evolutionary Integral Equations and Applications", Birkhäuser, Basel, 1993.
J. Pruss, Bounded solutions of Volterra equations, *SIAM Math. Anal.* **19**(1987), 133-149.
A. S. Rao, On almost automorphic solutions of certain abstract differential equations, , *Indian J. Math.* **33(2)** (1991),179-187.
S. Schweiker, "Mild Solutions of Second-Order Differential Equations on the Line", Math. Proc. Cambridge Philos. Soc. 129 (2000), no. 1, 129-151.
E. Schuler and Vu Quoc Phong, "The Operator Equation $AX - X\mathcal{D}^2 = -\delta_0$

and Second-Order Differential Equations in Banach Spaces", *"Semigroups of operators: theory and applications"* (Newport Beach, CA, 1998), 352-363, Progr. Nonlinear Differential Equations Appl., 42, Birkauser, Basel, 2000.

J.S. Shin, T. Naito, Semi-Fredholm operators and periodic solutions for linear functional differential equations, *J. Diff. Eq.* **153**(1999), 407-441.

G.M. Skylar, V.Ya. Shirman, On the asymptotic stability of a linear differential equation in a Banach space, *Teor. Funktsii Funktsional Anal. Prilozhen* **37** (1982), 127-132.

V.E. Sljusarcuk, Estimates of spectra and the invertibility of functional operators. (Russian) *Mat. Sb. (N.S.)* **105**(147) (1978), no. 2, 269–285.

D. Smart, *Fixed Point Theorems*, Cambridge University Press, Cambridge, 1980.

H. Tanabe, "Equations of evolution". Pitman , Boston, Mass.-London, 1979.

C.C. Travis, G.F. Webb, Existence and stability for partial functional differential equations, *Trans. Amer. Math. Soc.*, **200**(1974), 394-418.

Q.P. Vu, Stability and almost periodic of trajectories of periodic processes, *J. Diff. Eq.* **115**(1995), 402-415.

Q.P. Vu, Almost periodic solutions of Volterra equations, *Diff. Int. Eq.* **7**(1994), 1083-1093.

Q.P. Vu, E. Schuler, The operator equation $AX - XB = C$, stability and asymptotic behaviour of differential equations, *J. Diff. Eq.* **145** (1998), 394-419.

L. Weis, *Operator-valued Fourier multiplier theorems and maximal L_p-regularity.* Math. Ann. **319** (2001), 735-758.

J. Wu, "Theory and Applications of Partial Functional Differential Equations ", Applied Math. Sci. **119**, Springer, Berlin- New york, 1996.

T. Yoshizawa, "Stability Theory and the Existence of Periodic Solutions and Almost Periodic Solutions", Applied Math. Sciences 14, Springer-Verlag, New York, 1975.

S. Zaidman, "Topics in abstract differential equations", Pitman Research Notes in Mathematics Series, **304**, Longman Scientific & Technical, New York, 1994.

V. V. Zikov, Some questions of admissibility and dichotomy. The averaging principle. (Russian) *Izv. Akad. Nauk SSSR Ser. Mat.* **40** (1976), no. 6, 1380–1408,

Index